Global Perspectives on the Health of Seeds and Plant Propagation Material

Plant Pathology in the 21st Century

More information about this series at
http://www.springer.com/series/8169

Maria Lodovica Gullino • Gary Munkvold
Editors

Global Perspectives on the Health of Seeds and Plant Propagation Material

Volume 6

Springer

Editors
Maria Lodovica Gullino
AGROINNOVA
Università degli Studi di Torino
Grugliasco
Torino
Italy

Gary Munkvold
Plant Pathology and Microbiology
Iowa State University
Ames
Iowa
USA

ISBN 978-94-017-9388-9 ISBN 978-94-017-9389-6 (eBook)
DOI 10.1007/978-94-017-9389-6
Springer Dordrecht Heidelberg New York London

Library of Congress Control Number: 2014952658

Printed on acid-free paper

Springer is part of Springer Science+Business Media (www.springer.com)

Foreword

This volume continues the series of books on "Plant Pathology in the 21st Century", which started in 2010, in cooperation with the International Society for Plant Pathology and contains the papers given at the 10th International Congress of Plant Pathology (ICPP 2013) held in Beijing, August 25–30, 2013 concerning seed health.

The use of healthy seeds and propagation material is a prerequisite in any cropping systems, because it permits to strongly reduce the further adoption of other disease management strategies in the field during the cultivation.

Many pathogens are transmitted throughout infected seeds and propagation material. The fact that propagation material production is very much concentrated in few establishments, favors the quick spread of new diseases throughout seed commercialization. This phenomenon is very much accelerated in a globalized system.

The book covers case studies of contamination, aspects of detection and diagnosis as well as disease management strategies, with special emphasis towards seed treatments with unconventional products.

We believe that, besides representing a written testimony of ICPP 2013, this book will be useful for all plant pathologists as well as students in advanced courses.

We wish to thank all the colleagues who accepted to be part of this book, Zuzana Bernhart and her group at Springer for their continuous support and Laura Castellani for her skilfull technical assistance.

The Editors

Contents

Contributors

Theresa A. S. Aveling University of Pretoria, Pretoria, South Africa

Giorgio M. Balestra Department of Agriculture, Forestry, Nature and Energy, University of Tuscia, Tuscia, Italy

Jianjun Feng Animal & Plant Quarantine Technology Center, Shenzhen, People's Republic of China

Angelo Garibaldi AGROINNOVA, University of Torino, Grugliasco, Italy

Giovanna Gilardi AGROINNOVA, University of Torino, Grugliasco, Italy

Daminen Griessinger European and Mediterranean Plant Protection Organization, Paris, France

M. Lodovica Gullino AGROINNOVA, DISAFA, University of Torino, Grugliasco, Italy

Eckhard Koch Julius Kühn-Institute, Dresden, Germany

Jianqiang Li Department of Plant Pathology, China Agricultural University, Beijing, People's Republic of China

Laixin Luo Department of Plant Pathology, China Agricultural University, Beijing, People's Republic of China

Angelo Mazzaglia Department of Agriculture, Forestry, Nature and Energy, University of Tuscia, Tuscia, Italy

Madeleine McMullen European and Mediterranean Plant Protection Organization, Paris, France

Carmen N. Mortensen Danish Seed Health Centre, Department of Plant and Environmental Sciences, University of Copenhagen, Copenhagen, Denmark

Gary P. Munkvold Seed Science Center, Iowa State University, Ames, IA, USA

Gilberto Olaya Syngenta Crop Protection, Vero Beach, FL, USA

Françoise Petter European and Mediterranean Plant Protection Organization (EPPO), Paris, France

Angelo Porta-Puglia Istituto Sperimentale per la Patologia Vegetale (presently CRA-PAV), Rome, Italy

Steven J. Roberts Plant Health Solutions Ltd, Warwick, UK

Anne-Sophie Roy European and Mediterranean Plant Protection Organization, Paris, France

Sabrina Sarrocco Dipartimento di Scienze Agrarie, Alimentari e Agro-ambientali, Università di Pisa, Pisa, Italy

Norman W. Schaad USDA, ARS, Foreign Disease Weed Science-RU, Ft. Detrick, MD, USA

Thousand Oaks, CA, USA

Monika Scheller Syngenta Crop Protection, Basel, Switzerland

Heather E. Simmons Syngenta Crop Protection, Basel, Switzerland

Muriel Suffert European and Mediterranean Plant Protection Organization, Paris, France

Giovanni Vannacci Dipartimento di Scienze Agrarie, Alimentari e Agro-ambientali, Università di Pisa, Pisa, Italy

Clifford Watrin Syngenta Crop Protection AG, Basel, Switzerland

Ronald Zeun Syngenta Crop Protection, Stein, Switzerland

Part I
General Aspects

Chapter 1
Seed Transmission in the *Potyviridae*

Heather E. Simmons and Gary P. Munkvold

Abstract Viral pathogens comprise approximately half of the emerging diseases in plants, and plant introductions (including the international movement of seed) are considered to be one of the most important contributing factors to the emergence of these pathogens. For the most part plant viruses are incapable of surviving outside of host tissue making their long-term propagation dependent on their hosts. Thus infected seeds are an effective strategy that not only allows for pathogen survival from one season to the next, but also for their dispersal. The *Potyviridae,* as the largest plant virus family, is often considered to be the most economically important and its members rank among the most successful plant pathogens. Seed transmission within the *Potyviridae* family is not uncommon, however the exact mechanism of viral entry into the germ line is currently unknown, and the genetic basis of seed transmission has yet to be completely elucidated. Seed transmission rates are influenced by complex interactions among a variety of factors including the host cultivar, the virus isolate, environmental conditions, the timing of infection, vector characteristics, and viral synergism. Seed transmission can have an enormous effect on the epidemiology of crop pathogens due in part to the ecology of plant viruses which are often secondarily disseminated via insect vectors with the effect that extremely low frequencies of seed transmission can result in devastating epidemics. This is compounded by the fact that vertically infected seedlings often do not exhibit symptoms of viral infection. Given the potential for seed transmitted viral pathogens to initiate epidemics, it is vital to understand how seed transmission rates translate into epidemics. In addition, as seed transmission is a means of dispersal for these viral pathogens, effective phytosanitary measures to control the spread of these pathogens are crucial.

Keywords Epidemiology • Seed infection • Seed-to-seedling transmission • *Potyviridae* • Virus

Given that approximately 90 % of the food crops grown worldwide are propagated from seed (Maude 1996) it is hardly surprising that seed transmitted pathogens

H.E. Simmons • G.P. Munkvold (✉)
Seed Science Center, Iowa State University, Ames, IA 50010, USA
e-mail: munkvold@iastate.edu

M.L. Gullino, G. Munkvold (eds.), *Global Perspectives on the Health of Seeds and Plant Propagation Material*, Plant Pathology in the 21st Century 6,
DOI 10.1007/978-94-017-9389-6_1

would be a significant concern for both growers and industry alike. Seed transmission is an effective strategy for pathogens, especially viruses, to maintain their populations in host plants. In 1972, K.F. Baker wrote, "Seed transmission is now recognized as the method *par excellence* by which plant pathogens (a) are introduced into new areas, (b) survive periods when the host is lacking, (c) are selected and disseminated as host-specific strains, and (d) are distributed through the plant population as foci of infection" (Baker 1972). Most viruses are unable to survive for any length of time outside host issue, making long-term perpetuation of viruses particularly difficult, especially for those that infect annual plants. Seed infection is an effective mechanism to overcome this, so that the long-term survival of the pathogen is linked to the host (Stacie-Smith and Hamilton 1988). This mechanism allows not only for the survival of the pathogen from one season to the next, but also for the long distance dissemination of the pathogen via infected seed (Albrechtsen 2006). One such example is *Wheat streak mosaic virus* (WSMV) for which phylogenetic studies suggest that the introduction of this virus and its subsequent distribution within Australia was likely via imported seed (Dwyer et al. 2007).

Approximately 20 % of all plant viruses are seed transmitted (Mink 1993), and it is believed that approximately one third of plant viruses will eventually be shown to be seed transmitted (Stacie-Smith and Hamilton 1988). Currently 231 viruses are believed to be seed transmitted (Sastry 2013), with 13 % of these being members of the *Potyviridae* (See Table 1.1 for a list of seed transmitted *Potyviridae*). Among the viruses that infect plants the *Potyviridae* is the largest family, and as a result are often considered to be the most economically important (Berger 2001). This family, and in particular the aphid-transmitted members, are among the most successful plant pathogens (Rybicki and Pietersen 1999). Some of the most important crop pathogens are members of the *Potyviridae*, including *Bean common mosaic virus* (BCMV), *Maize dwarf mosaic virus*, *Lettuce mosaic virus* (LMV), *Plum pox virus* (PPV), *Potato virus Y* (PVY), WSMV, and *Zucchini yellow mosaic virus* (ZYMV) (Berger 2001).

The *Potyviridae* is composed of eight genera: *Brambyvirus*, *Bymovirus*, *Ipomovirus*, *Macluravirus*, *Poacevirus*, *Potyvirus*, *Rymovirus*, and *Tritimovirus*. In addition, there is one as yet unassigned group, which consists of two viruses (*Spartina mottle virus* and *Tomato mild mottle virus*). These genera have a combined total of 203 species with the *Potyvirus* group being the largest, comprising 146 members (International Committee on Taxonomy of Viruses, 2012). The classification is based on shared characteristics; all have positive sense RNA genomes, all save one (*Bymovirus*) are monopartite, and they share a gene order as well as sequence homology. The genomes of all members have a VPg (viral protein genome-linked) covalently linked to the 5′ end and a polyadenylated 3′ end. They also all share the presence of the distinctive pinwheel inclusion bodies of the Cylindrical Inclusion (CI) protein. Genera and species are differentiated based on sequence identity, host range, transmission mode, cytopathology, vector transmission and antigenic properties (King et al. 2012; López-Moya et al. 2001).

Averaged estimates in the late 1990s of worldwide crop losses due to viruses were between 1 and 7 % depending on the crop species (Oerke and Dehne 2004).

Table 1.1 Seed transmitted *Potyviridae*, their acronyms, and their important hosts[a]

Virus	Acronym	Important Host (Genus)
Artichoke latent virus	ArLV	*Cynara*
Bean common mosaic	BCMV	*Phaseolus, Vigna*
Bean yellow mosaic	BYMV	*Lupinus, Vicia, Pisum, Melilotus*
Blackeye cowpea mosaic	BICM	*Vigna*
Cassia yellow spot	CasYSV	*Cassia*
Cowpea aphid-borne Mosaic	CABMV	*Glycine, Phaseolus*
Cowpea green vein banding virus	CGVBV	*Phaseolus, Vigna*
Desmodium mosaic	DesMV	*Desmodium*
Guar symptomless	GSLV	*Cyamopsis*
Hippeastrum mosaic	HiMV	*Hippeastrum*
Leek yellow stripe	LYSV	*Allium*
Lettuce mosaic	LMV	*Lactuca, Senecio*
Maize dwarf mosaic	MDMV	*Zea*
Mungbean mosaic	MbMV	*Vigna*
Onion yellow dwarf	OYDV	*Allium*
Papaya ringspot	PRSV	*Carica*
Pea seedborne mosaic	PSbMV	*Pisum*
Peanut mottle	PeMoV	*Arachis, Glycine, Vigna, Voandzeia*
Peanut stripe	PStV	*Arachis, Glycine, Vigna*
Plum pox	PPV	*Prunus*
Potato Y	PVY	*Solanum*
Soybean mosaic	SMV	*Glycine, Lupinus, Phaseolus*
Sugarcane mosaic	SCMV	*Zea*
Sunflower mosaic	SuMV	*Helianthus*
Telfairia mosaic	TeMV	*Telfairia*
Tobacco etch	TEV	*Nicotiana, Solanum, Capsicum*
Turnip mosaic	TuMV	*Raphanus*
Watermelon mosaic	WMV	*Cucumis, Echinocystis*
Wheat streak mosaic	WSMV	*Triticum*
Zucchini yellow mosaic	ZYMV	*Cucurbita, Ranunculus*

Adapted from Singh and Mathur (2004). Data from Mink (1993), Sastry (2013) and Albrechtsen (2006)

However the economic impact of individual viral epidemics can be enormous. For example, yield losses of up to 98.7 % have been reported for WSMV (Edwards and Mcmullen 1988), and annual losses in Kansas alone due to WSMV have exceeded $30 million (Jons et al. 1981). PVY yield losses range from 10 % and 80 % (Valkonen 2007), and the global economic impact of PPV over a 20 year period is estimated to be 576 million Euros (Cambra et al. 2006). Given the current increase in emerging pathogens that is occurring these figures are likely to increase. Emerging viral pathogens are significant and constitute 47 % of emerging diseases in plants, with plant introductions (including the international movement of seed) being thought to be one of the most important contributing factors to their emergence (Anderson et al. 2004). Other reasons for this increase include conversion of natural vegetation to agriculture, climate change as well as an expansion in trade and globalization (Jones 2009). Global warming is likely to affect the rate at which plant RNA viruses evolve as RNA replication is affected by temperature as are plant defenses (Elena 2011). In addition, climate change will undoubtedly influence the geographical distribution of crops and plants in natural ecosystems, and by extension their pathogens and vectors. Warming trends are expected to change the distribution, winter survival and spring arrival of insect vectors (Anderson et al. 2004), potentially affecting viral epidemics, and this has already been observed, for example, with *Barley yellow mosaic virus* (Coakley et al. 1999). The human population is estimated to reach nine billion by 2050 (Cohen 2003), and the Food and Agriculture Organization estimates that global food production will need to increase by 60 % by 2050 (Alexandratos and Bruinsma 2012). As agricultural intensification is thought to expedite the establishment and spread of emerging viruses (Elena 2011) it is extremely likely that we will continue to see an increase in emerging viral pathogens. Given that seed transmission is instrumental in the epidemiology of viral diseases, as it serves as a means of dispersal, both as seeds and as an initial source of infection for vector dispersal (Mink 1993), the need to establish rigorous phytosanitary measures for these viral pathogens will become increasing important.

1 Mechanisms of Seed Transmission

Although seed transmission within the *Potyviridae* family is not unusual, the mechanism by which the virus enters the germ line is currently unknown. However, two possible routes of embryonic infection have been postulated: direct invasion of the embryonic tissue after fertilization, or infection of the gametes prior to fertilization, either through the ovules or via pollen. In addition, it has been suggested that the seed transmission rate may be a sum of both indirect and direct embryotic invasion (Wang and Maule 1994). Plant viruses differ from animal viruses in the sense that movement of animal viruses into the cell is via receptor-mediated mechanisms, with the effect that these viruses can exploit the extracellular environment. Plant viruses, in contrast, are restricted to the intracellular compartments of the host and cell-to-cell movement is regulated by the plasmodesmata (Maule and Wang 1996). Several viral proteins are involved in cell-to-cell movement and it is thought that the Coat Protein (CP) binds to the viral RNA and alters the exclusion size limit of the plasmodesmata. This phenomenon is thought to follow the infection front and is transient (Heinlein et al. 1995; Oparka et al. 1997). The helper component protein (HC-Pro) is thought to increase plasmodesmal permeability (Rojas et al. 1997), and the CI is believed to guide the CP-RNA complex to the plasmodesmata (Rodriguez-Cerezo et al. 1997). In order for systemic infection to occur, the virus must enter the vascular tissue. The virus moves from the mesophyll cells and through a series of cells, which are the perivascular parenchyma, the phloem parenchyma, the companion cells, and finally into the sieve tube elements (Astier et al. 2001). In the *Potyviridae* the CP is necessary for viral movement within host plant tissues, and it is thought that the HC-Pro functions in aiding the entry and exit of the virus into and out of the host vascular system (Urcuqui-Inchima et al. 2001). Viral movement through the plant is directed in the sense that it moves with the carbon metabolites that are transported from the source leaves to the sink immature leaves; in other words, viral movement follows the same path as the photoassimilates (Maule and Wang 1996).

The viral genetic basis of seed transmission has yet to be completely determined; however it appears that a number of viral genes are involved in seed transmission. Chimeras of transmissible and nontransmissible strains of *Pea seedborne mosaic virus* (PSbMV) revealed that the 5′ untranslated region (UTR), the HC-Pro and the CP region of the potyvirus genome may be involved in the seed transmission of this virus (Johansen et al. 1996). The CI may also be involved in seed transmission and in PSbMV infections; cylindrical inclusions were observed over plasmodesmatal openings at the testa-endosperm boundary wall (Roberts et al. 2003). Notably all of the proteins thought to be involved in seed transmission are also involved in viral movement save one (the 5′UTR). When considering seed transmission, the mode of virus movement within the plant will have an enormous effect on the potential for vertical transmission, and phloem limited viruses are generally not seed transmissible (Mink 1993).

Evidence for the direct invasion of the embryo is derived from work with PSbMV. There is some evidence in PSbMV that the virus uses the suspensor as a mode of entry into the embryonic tissues. After fertilization, the zygote undergoes an asymmetrical cell division, resulting in a small apical cell, which will become the embryo and a larger basal cell (the suspensor). In pea the suspensor provides nutrients for the growing embryo from the endosperm and appears to be anchored close to the micropyle (a tiny opening in the ovule through which the pollen tube enters) during the early stages of seed development (Wang and Maule 1994). It is believed that embryonic invasion occurs as a result of viral movement from the maternal cells in the micropyle to the endospermic cytoplasm and embryonic suspensor from where it invades the embryo (Roberts et al. 2003). Given that the embryonic suspensor undergoes a programmed cell death, the ability of the virus to gain entry into the embryo in this manner is transient (Wang and Maule 1994), and thus it appears that seed transmission of viral pathogens in this manner is dependent at least partially on timing and/or chance (Roberts et al. 2003).

Seed infection can occur via pollen although the frequency of transmission to seedlings through pollen is generally thought to be less than through the ovules (Mink 1993). Evidence for the indirect invasion of the embryo via the ovules is fairly extensive and has been demonstrated for viruses in families other than the *Potyviridae* (e.g., *Tobacco ringspot virus* (*Secoviridae*) (Yang and Hamilton 1974), *Barley stripe mosaic virus* (*Vigraviridae*) (Carroll and Mayhew 1976), *Cucumber Mosaic virus* (*Bromoviridae*) (Yang et al. 1997), and *Turnip yellow mosaic* (*Tymoviridae*) (de Assis Filho and Sherwood 2000). With respect to the *Potyviridae* the evidence for seed transmission via the ovules is substantially less than for other viral families and in some instances is contradictory, for example, in LMV, there are studies indicating that the seed transmission of this virus does occur through the ovules (Ryder 1964), and others reporting that it does not (Hunter and Bowyer 1994). However there is some evidence of indirect invasion via pollen in a number of *Potyviridae* members, for example, in LMV (Hunter and Bowyer 1997; Ryder 1964) using electron microscopy and immunogold labeling LMV was found to infect the pollen mother cells (Hunter and Bowyer 1997). Serological work with PSbMV revealed that the virus was present in pollen in seed transmissible variants but was absent from these tissues in non seed transmissible isolates (Kohnen et al. 1995). For some viruses, such as BCMV, there is evidence that direct invasion of the embryo may occur through both the pollen and the ovules (Schippers 1963; Medina and Grogan 1961). The seed to seedling transmission rate of *Sugarcane mosaic virus* (SCMV) is postulated to be a sum of the direct invasion of the embryo and indirect invasion of infected pollen. The overall seed to seedling transmission rate of SCMV in maize was determined to be 4.81 % and the rate of transmission as a result of infected pollen grains was 0.04–0.10 % (Li et al. 2007).

There is a distinction between viruses that infect the embryo versus those that are found in other seed tissues or remain as contaminants on the seed surface. This is significant because embryo infection can readily result in seedling infection while viral infection of other seed parts will only result in seedling infection if the virus is easily transmitted mechanically and is resistant to inactivation. Viruses present on or near the seed surface are often eliminated by heat or chemical treatments and this is in direct contrast to embryonic infection where the treatments to inactivate the virus would potentially also kill the embryo (Stacie-Smith and Hamilton 1988). Seed transmission as a result of the virus being carried on the seed surface is fairly uncommon (Johansen et al. 1994) and has not been demonstrated in the *Potyviridae*. Examples of this mode of transmission include members of the tabomoviruses, with the only outside member being *Southern bean mosaic virus* (*Sobemovirus*). It would thus appear that the bulk of viruses that are seed-transmitted are carried within the embryo (Albrechtsen 2006).

2 Factors That Influence Seed Transmission

There are a number of factors that influence seed transmission rates and these include the host cultivar, the virus isolate, interactions with the environment, the timing of infection and viral synergism. It has been postulated that whether or not seed transmission occurs is primarily affected by host-virus interactions and the timing of infection with environment playing a lesser role (Mink 1993). In addition, the relation of the virus to its vector may have an effect on seed transmission and viruses that are horizontally transmitted in a persistent manner are typically not seed transmitted whereas those transmitted nonpersistently tend to be seed transmitted (Bennett 1969).

Different cultivars within a species can vary in their seed transmission rates. For instance, using LMV the incidence of seed transmission ranged from 1 % to 8 % depending on the variety of lettuce (Grogan and Bardin 1950). Similarly an investigation of seed transmission of PSbMV in 38 pea cultivars revealed that five of these exhibited no seed transmission whatsoever (Stevenson and Hagedorn 1973). This varietal variation in seed transmission may have a genetic basis and in soybean seed transmission of *Soybean mosaic virus* (SMV) appears to be a polygenic trait and a number of genes are necessary for high rates of transmission (Domier et al. 2011). Transmission of different isolates of the same virus can also vary within a single host. For instance work with *Peanut mottle mosaic virus* (PeMoV) revealed differences in the frequency of seed transmission as a function of virus isolate (Adams and Kuhn 1977). Similarly within PSbMV there are both transmissible and nontransmissible isolates (Roberts et al. 2003). An investigation of 14 bean cultivars and four virus isolates showed that seed transmission of BCMV was influenced by both isolate strain as well as the host cultivar (Morales and Castano 1987). Seed transmission rates may also be influenced by the interaction of host cultivar and virus isolate. An investigation of eight soybean cultivars and seven SMV isolates (Tu 1989) found that the interaction between host cultivar and virus isolate resulted in the seed transmission rate varying from zero to 70 %; a resistant cultivar had overall lower seed transmission than the susceptible cultivar, but at least one SMV strain was seed-transmitted at a higher rate in the resistant cultivar.

The environment can affect seed transmission rates and studies using SMV elucidated that temperature had an effect on seed transmission in soybean. Although virus symptoms on the mother plants were most severe when plants were grown at 25 °C, seed transmission was optimal when the plants were grown at 20 °C (average 48 %) and seed transmission decreased at 15 °C (average 7 %) and 25 °C (average 9.7 %) (Tu 1992). Work with PSbMV determined that reduced rainfall decreased the incidence of virus in the field because it resulted in a delay of the vector (Coutts et al. 2009). Thus the risk associated with a given level of seed infection was dependent on conditions before and after planting. It is apparent that the factors influencing seed transmission rates are complex, as suggested by Maule and Wang (1996), and are the result of multifaceted interactions between the host, virus, vector and environment.

The seed transmission rate can be greatly influenced by the age of the host (flowering) at the time of inoculation. Seed transmission rates appear to be inversely related to the age of the plant (and/or developmental stage) upon infection (Wang and Maule 1992). In SMV a reduction in seed transmission of 13 % (16–3 %) was seen after the onset of flowering (Bowers and Goodman 1979). Likewise the date of inoculation was seen to influence the incidence of seed transmission in BCMV with the effect that seed transmission increased significantly if inoculation occurred within the first 20 days of the vegetative period of the host. In the same study, only 2 of 14 bean cultivars exhibited seed transmission if inoculation occurred more than 30 days after sowing (Morales and Castano 1987). Other *Potyviridae* for which the age of the host appears to affect seed transmission rates include PSbMV (Wang and Maule 1992), BCMV (Kaiser and Mossaheb 1974) and PeMoV (Paguio and Kuhn 1974).

Synergism can affect seed transmission rates although the direction of influence appears to vary, for instance, co-infections of PSbMV with *Pea early browning virus* (PEBV) (*Virgaviridae*) resulted in seed transmission being blocked in PSbMV although it was unaffected in PEBV (Wang and Maule 1997). For viruses in other families a synergistic effect that increases the rate of seed transmission has been reported e.g. *Turnip yellow mosaic virus* (*Tymoviridae*) (de Assis Filho and Sherwood 2000), and *Southern bean mosaic virus* (Sobemovirus) (Kuhn and Dawson 1973).

3 Seed to Seedling Transmission

Although the majority of seed transmission events require embryonic infection, embryo infection itself does not necessarily result in seedling infection. In fact the discrepancy between seed infection rates and seed to seedling transmission rates can vary greatly. For instance in ZYMV the seed infection rate was significantly higher (21.9 %) than the seed to seedling transmission rate (1.8 %) (Simmons et al. 2013). Similar results were found with LMV (Hunter and Bowyer 1993). However, there are instances where the whole seed-assay matches up with the seed transmission rate, for example in SMV (Bowers and Goodman 1979) and *Peanut stripe virus* (Xu et al. 1991), but this does not appear to be the norm (Albrechtsen 2006). One possible explanation for this is that inactivated viruses can occur in parts of the seed other than the embryo, with the effect that the virus is still detectable via serological or molecular methods. In these instances testing the whole seed for the virus will lead to an overestimation of the actual seed transmission rate. For viruses that infect the embryo, the seed to seedling infection rate is going to be the result of two factors: the first is the ability of the virus to survive in the embryo, and the second is its ability to be reactivated (Albrechtsen 2006). It is believed that a virus that has infected the embryo will remain viable for as long as the seed is viable (Bennett 1969) and there are examples of extreme longevity of seeds and their pathogens; BCMV has been shown to be able to survive and remain infectious for 30 years in seed (Pierce and Hungerford 1929).

Symptoms in seedlings are variable and appear to be dependent on the virus strain, host genotype and environment (Albrechtsen 2006). For example, with BCMV in some seedlings, viral symptoms did not appear until the second or third trifoliate leaf (Kaiser and Mossaheb 1974). Vertically infected seedlings often exhibit little to no symptoms of viral infection (Stacie-Smith and Hamilton 1988), and as a result visual inspection is frequently not the optimal method for determining the incidence of seed transmission for these pathogens. Cucurbit seedlings vertically infected with ZYMV demonstrated little to no visual symptoms (Simmons et al. 2013; Muller et al. 2006; Gleason and Provvidenti 1990), while slight symptoms have been observed in PSbMV (Hampton 1972). There may be a genetic basis for this and Illumina sequencing of ZYMV populations revealed that the 5′ UTR is highly variable in the seed transmitted populations compared to those transmitted horizontally. In this example the vertically transmitted populations were symptomless in comparison to the horizontally transmitted populations (Simmons et al. 2013). Likewise, studies with PPV determined that a deletion in this region resulted in reduced symptom development (Simon-Buela et al. 1997) and in BCMV an insertion in this region resulted in an increase in symptom severity (Zheng et al. 2002).

For some members of the *Potyviridae* the presence of the virus in the seed does not appear to affect germination rates, for instance in PSbMV (Hampton 1972), and in BCMV (Raizada et al. 1990; Hao et al. 2003). For others, however, there does appear to be an interaction between seed infection and low germination rate. This could potentially lessen the effect of epidemics, as only a subset of virally infected seeds will successfully initiate infections in subsequent generations. For instance in ZYMV the germination rate of seeds extracted from fruits from infected parents was 22.5 % versus 87.5 % for those harvested from non-infected parents (Simmons et al. 2013). This could be due to a number of reasons. It is possible that the lower germination rate could be the result of the effects of the pathogen on the mother plant, or the virus could be reducing the viability of the seed. This was found with SMV where infection severely reduced the seed yield. Viral infection reduced seed yield on average 58.5 % among eight soybean cultivars inoculated with seven virus isolates (Tu 1989). Similarly low numbers of viable seed were reported from ZYMV infected plants (Desbiez and Lecoq 1997). It is also possible that the viral titers in the seeds are simply too low to consistently initiate effective infections in the subsequent generation. A determination of viral titers via qPCR revealed that the titers of ZYMV were several orders of magnitude lower in the seed (11.3–60 ng/μl) than in the leaf (2,000–3,400 ng/μl) (Simmons et al. 2013). Alternatively, the viral population may be severely constrained by the host plant such that only a subset of the viral population is transmitted from the seed to the seedling, or host defense mechanisms, such as RNA silencing, may be eliminating the viral population, or preventing it from being transmitted to the seedling. An investigation of the numbers of infectious PSbMV particles that were subsequently transmitted to a vertically infected seedling from the mother plant was on average only one, suggesting that the bottleneck for this mode of transmission is very severe indeed (Fabre et al. 2014).

The ecology of plant viruses is such that seed transmission can have an enormous effect on the epidemiology of crop pathogens. This is due to the fact that the majority of plant viruses are secondarily disseminated via insect vectors; therefore small initial numbers of infected plants can lead to damaging epidemics (Maule and Wang 1996). This is particularly important as many of the plant viral vectors transmit nonpersistently, which means that insecticides are not effective at suppressing secondary spread. This is due to the fact that both acquisition and inoculation occur rapidly (within a few seconds), and thus the vector is not exposed for a sufficiently long enough period of time for the pesticide to be effective in reducing viral spread. As a result, the vector can often spread the virus to a neighboring plant before it is negatively affected by an insecticide (Perring et al. 1999). In addition the frequency at which seed transmission occurs may not be a good predictor of the epidemiological significance of a virus, and even extremely low transmission rates can initiate severe epidemics. For example LMV at an incidence of 0.001 resulted in an epidemic as a result of secondary spread via the insect vector (Ryder 1973). Similarly PeMoV at a seed transmission of 0.1 % is sufficient in the epidemiology of this pathogen (Adams and Kuhn 1977).

Given the potential for seed transmitted viral pathogens to initiate epidemics and the fact that viral pathogens are the most abundant group of emerging pathogens in plants, it is vital to understand how seed transmission rates translate into epidemics. There has been a lack of research examining the relationship between seed infection levels on the development of viral epidemics and the resulting risk to crop yields. This is vitally important for informing both seed industry as well as farmers (Jones 2000). Except for a few notable *Potyviridae* examples such as PSbMV, where research resulted in a threshold value of >0.5 % seed infection (Coutts et al. 2009), and LMV, for which a threshold of 0.1 % was established (Tomlinson 1962; Zink et al. 1956) very little quantitative work has been performed in this area. Although not *Potyviridae* members an excellent example of determining threshold levels was undertaken with two *Bromoviridae* (*Cucumber mosaic virus* and *Alfalfa mosaic virus*). This study underscores the necessity of determining threshold levels that are based on an examination of particular host-virus interactions, specific geographic sites as well as year-to-year variations (Jones 2000). Given that different viral pathogens will have different thresholds it is critical that this type of research be conducted on individual virus-host pathosystems, as generalities cannot be applied to specific cases.

References

Adams DB, Kuhn CW (1977) Seed transmission of peanut mottle virus in peanuts. Phytopathology 67(9):1126–1129

Albrechtsen SE (2006) Testing methods for seed-transmitted viruses: principles and protocols. CABI Publishing, Wallingford

Alexandratos N, Bruinsma J (2012) World agriculture towards 2030/2050: the 2012 revision. ESA Working paper

Anderson PK, Cunningham AA, Patel NG, Morales FJ, Epstein PR, Daszak P (2004) Emerging infectious diseases of plants: pathogen pollution, climate change and agrotechnology drivers. Trends Ecol Evol 19(10):535–544

Astier S, Albouy J, Maury Y, Lecoq H (2001) Principles of plant virology: genome, pathogenicity, virus ecology. Institut National de la Recherche Agronomique, Paris, France

Baker KF (1972) Seed pathology. In: Kozlowski TT (ed) Germination control. Metabolism, and pathology. Academic, New York, pp 317–416

Bennett CW (1969) Seed transmission of plant viruses. Adv Virus Res 14:221–261

Berger PH (2001) Potyviridae. In: eLS. Wiley, Hoboken NJ

Bowers G Jr, Goodman R (1979) Soybean mosaic virus: infection of soybean seed parts and seed transmission. Phytopathology 69(6):569–572

Cambra M, Capote N, Myrta A, Llácer G (2006) Plum pox virus and the estimated costs associated with sharka disease. EPPO Bull 36(2):202–204

Carroll TW, Mayhew DE (1976) Occurrence of virions in developing ovules and embryo sacs of barley in relation to the seed transmissibility of barley stripe mosaic virus. Can J Bot 54 (21):2497–2512

Coakley SM, Scherm H, Chakraborty S (1999) Climate change and plant disease management. Annu Rev Phytopathol 37(1):399–426

Cohen JE (2003) Human population: the next half century. Science 302(5648):1172–1175

Coutts BA, Prince RT, Jones RAC (2009) Quantifying effects of seedborne inoculum on virus spread, yield losses, and seed infection in the pea seed-borne mosaic virus-field pea pathosystem. Phytopathology 99(10):1156–1167

de Assis Filho F, Sherwood J (2000) Evaluation of seed transmission of turnip yellow mosaic virus and tobacco mosaic virus in *Arabidopsis thaliana*. Phytopathology 90(11):1233–1238

Desbiez C, Lecoq H (1997) Zucchini yellow mosaic virus. Plant Pathol 46(6):809–829

Domier LL, Hobbs HA, McCoppin NK, Bowen CR, Steinlage TA, Chang S, Wang Y, Hartman GL (2011) Multiple loci condition seed transmission of soybean mosaic virus (SMV) and SMV-induced seed coat mottling in soybean. Phytopathology 101(6):750–756

Dwyer GI, Gibbs MJ, Gibbs AJ, Jones RAC (2007) Wheat streak mosaic virus in Australia: relationship to isolates from the pacific northwest of the USA and its dispersion via seed transmission. Plant Dis 91(2):164–170

Edwards MC, Mcmullen MP (1988) Variation in tolerance to wheat streak mosaic-virus among cultivars of hard red spring wheat. Plant Dis 72(8):705–707

Elena SF (2011) Evolutionary constraints on emergence of plant RNA viruses. Recent Adv Plant Virol 283–300

Fabre F, Moury B, Johansen EI, Simon V, Jacquemond M, Senoussi R (2014) Narrow bottlenecks affect pea seedborne mosaic virus populations during vertical seed transmission but not during leaf colonization. PLoS Pathog 10(1):e1003833

Gleason M, Provvidenti R (1990) Absence of transmission of zucchini yellow mosaic virus from seeds of pumpkin. Plant Dis 74(10)

Grogan RG, Bardin R (1950) Some aspects concerning seed transmission of lettuce mosaic virus. Phytopathology 40:965

Hampton RO (1972) Dynamics of symptom development of seed-borne pea fizzletop virus. Phytopathology 62(2):268

Hao NB, Albrechtsen SE, Nicolaisen M (2003) Detection and identification of the blackeye cowpea mosaic strain of bean common mosaic virus in seeds of *Vigna unguiculata* sspp. from North Vietnam. Australasian Plant Pathol 32(4):505–509

Heinlein M, Epel BL, Padgett HS, Beachy RN (1995) Interaction of tobamovirus movement proteins with the plant cytoskeleton. Science 270(5244):1983–1985

Hunter DG, Bowyer JW (1993) Cytopathology of lettuce mosaic-virus-infected lettuce seeds and seedlings. J Phytopathol 137(1):61–72

Hunter DG, Bowyer JW (1994) Cytopathology of mature ovaries from lettuce plants infected by lettuce mosaic Potyvirus. J Phytopathol Phytopathol Z 140(1):11–18

Hunter DG, Bowyer JW (1997) Cytopathology of developing anthers and pollen mother cells from lettuce plants infected by lettuce mosaic Potyvirus. J Phytopathol 145(11–12):521–524

Johansen E, Edwards MC, Hampton RO (1994) Seed transmission of viruses: current perspectives. Annu Rev Phytopathol 32:363–386

Johansen IE, Dougherty WG, Keller KE, Wang D, Hampton RO (1996) Multiple viral determinants affect seed transmission of pea seedborne mosaic virus in Pisum sativum. J Gen Virol 77 (12):3149–3154

Jones RAC (2000) Determining 'threshold' levels for seed-borne virus infection in seed stocks. Virus Res 71(1–2):171–183

Jones RAC (2009) Plant virus emergence and evolution: origins, new encounter scenarios, factors driving emergence, effects of changing world conditions, and prospects for control. Virus Res 141(2):113–130

Jons VL, Timian RG, Lamey HA (1981) Effect of wheat streak mosaic-virus on 12 hard red spring wheat cultivars. North Dakota Farm Res 39(2):17–18

Kaiser WJ, Mossaheb GH (1974) Natural infection of mungbean by bean common mosaic-virus. Phytopathology 64(9):1209–1214

King AMQ, Adams MJ, Carstens EB, Lefkowitz EJ (2012) Virus taxonomy: classification and nomenclature of viruses: ninth report of the international committee on taxonomy of viruses. Elsevier/Academic, London

Kohnen PD, Johansen IE, Hampton RO (1995) Characterization and molecular-detection of the P4 pathotype of pea seed-borne mosaic Potyvirus. Phytopathology 85(7):789–793

Kuhn CW, Dawson WO (1973) Multiplication and pathogenesis of cowpea chlorotic mottle virus and southern bean mosaic-virus in single and double infections in cowpea. Phytopathology 63 (11):1380–1385

Li L, Wang X, Zhou G (2007) Analyses of maize embryo invasion by Sugarcane mosaic virus. Plant Sci 172(1):131–138

López-Moya JJ, Valli A, García JA (2001) Potyviridae. In: eLS. Wiley, Hoboken NJ

Maude RB (1996) Seedborne diseases and their control: principles and practice. CAB International, Wallingford

Maule AJ, Wang D (1996) Seed transmission of plant viruses: a lesson in biological complexity. Trends Microbiol 4(4):153–158

Medina AC, Grogan RG (1961) Seed transmission of bean mosaic virus. Phytopathology 51:452–456

Mink GI (1993) Pollen-transmitted and seed-transmitted viruses and viroids. Annu Rev Phytopathol 31:375–402

Morales FJ, Castano M (1987) Seed transmission characteristics of selected bean common mosaic-virus strains in differential bean cultivars. Plant Dis 71(1):51–53

Muller C, Brother H, Bargen SV, Buttner C (2006) Zucchini yellow mosaic virus – incidence and sources of virus infection in field-grown cucumbers and pumpkins in the Spreewald, Germany. J Plant Dis Prot 113(6):252–258

Oerke EC, Dehne HW (2004) Safeguarding production – losses in major crops and the role of crop protection. Crop Prot 23(4):275–285

Oparka KJ, Prior DA, Santa Cruz S, Padgett HS, Beachy RN (1997) Gating of epidermal plasmodesmata is restricted to the leading edge of expanding infection sites of tobacco mosaic virus (TMV). Plant J 12(4):781–789

Paguio O, Kuhn C (1974) Incidence and source of inoculum of peanut mottle virus and its effect on peanut. Phytopathology 64(1):60–64

Perring TM, Gruenhagen NM, Farrar CA (1999) Management of plant viral diseases through chemical control of insect vectors. Annu Rev Entomol 44:457–481

Pierce WH, Hungerford CW (1929) A note on the longevity of the bean mosaic virus. Phytopathology 19(6):605–606

Raizada RK, Albrechtsen SE, Lange L (1990) Detection of bean common mosaic-virus in dissected portions of individual bean-seeds using immunosorbent electron-microscopy. Seed Sci Technol 18(3):559–565

Roberts IM, Wang D, Thomas CL, Maule AJ (2003) Pea seed-borne mosaic virus seed transmission exploits novel symplastic pathways to infect the pea embryo and is, in part, dependent upon chance. Protoplasma 222(1–2):31–43

Rodriguez-Cerezo E, Findlay K, Shaw JG, Lomonossoff GP, Qiu SG, Linstead P, Shanks M, Risco C (1997) The coat and cylindrical inclusion proteins of a potyvirus are associated with connections between plant cells. Virology 236(2):296–306

Rojas MR, Zerbini FM, Allison RF, Gilbertson RL, Lucas WJ (1997) Capsid protein and helper component proteinase function as potyvirus cell-to-cell movement proteins. Virology 237 (2):283–295

Rybicki EP, Pietersen G (1999) Plant virus disease problems in the developing world. Adv Virus Res 53:127–175

Ryder EJ (1964) Transmission of common lettuce mosaic virus through the gametes of the lettuce plant. Plant Dis Report 48:522–523

Ryder EJ (1973) Seed transmission of lettuce mosaic virus in mosaic resistant lettuce. J Am Soc Hortic Sci 98:610–614

Sastry KS (2013) Seed-borne plant virus diseases. Springer, New Delhi, pp 10–30

Schippers B (1963) Transmission of bean common mosaic virus by seed of *Phaseolus vulgaris* L. cultivar Beka. Acta Bot Neerlandica 12(4):433–497

Simmons HE, Dunham JP, Zinn KE, Munkvold GP, Holmes EC, Stephenson AG (2013) Zucchini yellow mosaic virus (ZYMV, Potyvirus): vertical transmission, seed infection and cryptic infections. Virus Res 176(1–2):259–264

Simon-Buela L, Guo HS, Garcia JA (1997) Long sequences in the 5′ noncoding region of plum Pox virus are not necessary for viral infectivity but contribute to viral competitiveness and pathogenesis. Virology 233(1):157–162

Singh D, Mathur SB (2004) Histopathology of seed-borne infections. CRC Press, Boca Raton, pp 200–203

Stacie-Smith R, Hamilton RI (1988) Inoculum thresholds of seedborne pathogens. Phytopathology 78:875–880

Stevenson WR, Hagedorn DJ (1973) Further studies on seed transmission of pea seedborne mosaic virus in *Pisum sativum*. Plant Dis Report 57(3):248–252

Tomlinson JA (1962) Control of lettuce mosaic by the use of healthy seed. Plant Pathol 11(2):61–64

Tu JC (1989) Effect of different strains of soybean mosaic virus on growth, maturity, yield, seed mottling and seed transmission in several soybean cultivars. J Phytopathol 126(3):231–236

Tu JC (1992) Symptom severity, yield, seed mottling and seed transmission of soybean mosaic-virus in susceptible and resistant soybean – the influence of infection stage and growth temperature. J Phytopathol Phytopathol Z 135(1):28–36

Urcuqui-Inchima S, Haenni AL, Bernardi F (2001) Potyvirus proteins: a wealth of functions. Virus Res 74(1–2):157–175

Valkonen JPT (2007) Chapter 28 – viruses: economical losses and biotechnological potential. In: Vreugdenhil D, Bradshaw J, Gebhardt C, Govers F, Mackerron DKL, Taylor MA, Ross HA (eds) Potato biology and biotechnology. Elsevier, Amsterdam, pp 619–641

Wang DW, Maule AJ (1992) Early embryo invasion as a determinant in pea of the seed transmission of pea seed-borne mosaic-virus. J Gen Virol 73:1615–1620

Wang DW, Maule AJ (1994) A model for seed transmission of a plant-virus – genetic and structural-analyses of pea embryo invasion by pea seed-borne mosaic-virus. Plant Cell 6 (6):777–787

Wang DW, Maule AJ (1997) Contrasting patterns in the spread of two seed-borne viruses in pea embryos. Plant J 11(6):1333–1340

Xu Z, Chen K, Zhang Z, Chen J (1991) Seed transmission of peanut stripe virus in peanut. Plant Dis 75:723–726

Yang AF, Hamilton R (1974) The mechanism of seed transmission of tobacco ringspot virus in soybean. Virology 62(1):26–37

Yang Y, Kim KS, Anderson EJ (1997) Seed transmission of cucumber mosaic virus in spinach. Phytopathology 87(9):924–931

Zheng H, Chen J, Chen J, Adams MJ, Hou M (2002) Bean common mosaic virus isolates causing different symptoms in asparagus bean in China differ greatly in the 5′-parts of their genomes. Arch Virol 147(6):1257–1262

Zink FW, Grogan RG, Welch JE (1956) The effect of the percentage of seed transmission upon subsequent spread of lettuce mosaic virus. Phytopathology 46(12):662–664

Chapter 2
Global Standards in Seed Health Testing

Theresa A.S. Aveling

Abstract Routine seed health testing is carried out in most countries for seed certification and plant quarantine. However, the majority of seed health tests used throughout the world have never been subject to rigorous validation. A fully validated test provides for analytical sensitivity, reproducibility and repeatability. Discrepancies between testing methods can occur, leading to costly phytosanitary disputes or liability claims. These issues can be avoided by working toward a system of universally accepted, standardized testing methods on a global level. To ensure that seed health tests are standardized and give reliable and reproducible results in accordance with the given specifications of the test methods, methods should go through a peer review system and/or collaborative study among laboratories. Three primary organizations publish standardized seed health tests: the International Seed Testing Association (ISTA), the International Seed Health Initiative (ISHI), and the U.S. National Seed Health System (NSHS). In 1957, the ISTA Plant Disease Committee (PDC) established a comparative seed health testing program to standardize techniques for detection of seed-borne pathogens. In 1993, the Seed Health Committee (SHC, formerly the PDC) began development of published guidelines for comparative testing. All ISTA validated methods are published in the International Rules for Seed Testing. Additionally, some ISHI-Veg methods have been accepted as ISTA Rules and as Standards by the NSHS. The procedures followed by ISTA, ISHI and the NSHS to achieve global standards in seed health testing are discussed.

Keywords International Seed Health Initiative • ISHI • International Seed Testing Association • ISTA • National Seed Health System • NSHS • Seed-borne pathogens • Validation

---

T.A.S. Aveling (✉)
University of Pretoria, Pretoria, South Africa
e-mail: Terry.Aveling@fabi.up.ac.za

M.L. Gullino, G. Munkvold (eds.), *Global Perspectives on the Health of Seeds and Plant Propagation Material*, Plant Pathology in the 21st Century 6,
DOI 10.1007/978-94-017-9389-6_2

1 Historical Review

Hiltner discussed methods for determining seed health of seed in germination tests at the second International Seed Testing Congress in Münster and Wageningen in 1910 (Mathur and Jørgensen 2002). However, according to Agarwal and Sinclair (1987), the first seed health testing laboratory in the world was only established in 1918 at the Government Seed Testing Station, Wageningen, the Netherlands with Lucie Doyer being appointed as the first official seed pathologist in 1919. The International Seed Testing Association (ISTA) was formed in 1924 at the fourth International Seed Testing Congress in Cambridge and at the next congress in 1928, L. Doyer reported on the activities of the "Plant Disease Committee" (PDC) which had split from the "Committee on Determinations of Genuineness of Variety in Field" established in 1924 (Mathur and Jørgensen 2002). The first recorded seed health methodologies were published by Dorogin in 1923 as *Instructions for testing seeds to determine the presence of fungus diseases at seed control stations* (Agarwal and Sinclair 1987) and Doyer in 1938 as the *Manual for the determination of seed-borne diseases* (Doyer 1938). In 1954 the European Plant Protection Organisation (EPPO), a regional organization established in accordance with the International Plant Protection Convention (IPPC) of 1951, sponsored by FAO, published a report of the first working party on seed-borne diseases titled *Danger from seed-borne diseases. A practical approach to the problem of international safeguard* (EPPO 1954). According to Mathur and Jørgensen (2002), the dominating factor influencing the development of seed health testing was the increased awareness of plant quarantine problems and extra pressure was put on ISTA to develop standard methods for such testing. A.J. Skolko, chairman of the ISTA PDC, presented the first report on comparative seed health testing of oats, flax, cabbage, wheat, barley and beet in 1956 at the 11th ISTA congress and the decision was made that the PDC would continue with comparative testing. P. Paul F.M. deNeergaard (Fig. 2.1), chairman of the ISTA PDC from 1956 to 1974, established the first comparative seed health testing programme to standardise

Fig. 2.1 P. Paul F.M. deNeergaard (Anon 1985)

techniques for the detection of seed-borne pathogens in 1957 (Mathur and Jørgensen 2002). This involved providing a number of independent seed scientists with referee seed samples and comparing results at an annual workshop of which the first was held in Cambridge, United Kingdom, in 1958 (Neergaard 1970). Neergaard played a leading role in the organisation of international cooperative testing of laboratory procedures for detection of pathogens on and in seed aiming at attaining uniformity and international standardisation (Anon 1985). In the last of three joint EPPO/ISTA Working Parties in London in July 1965, health certification of seeds for export was discussed and specified seed health testing methods developed and standardized by ISTA were recommended for quarantine inspection of seed lots (Neergaard 1970). Although the first ISTA *International Rules for Seed Testing* were published in 1931 (Steiner et al. 2008), the first specific seed health methods in the ISTA rules were only introduced in 1966 (Mathur and Jørgensen 2002; Muschick 2010). In 1998, Sheppard and Wesseling published a guide for comparative testing of methods for the detection of seed-borne pathogens. This was the basis for the *ISTA handbook of method validation for the detection of seed-borne pathogens* published by Sheppard and Cockerell (Fig. 2.2) in 2000. At that time Sheppard was the chairman of the then called Seed Health Committee (SHC). Based on this publication the Method Validation Working group was established in 2002 to develop the ISTA Method Validation Programme which came into force in 2007 and now applies to all seed quality testing (Hampton 2005, 2007). The term "method validation" is new in seed testing. According to Steiner et al. (2008) it was previously called "method standardization" or "method elaboration".

The International Seed Health Initiative (ISHI) also develops, evaluates and disseminates information on seed health test protocols. The International Seed Health Initiative for vegetables (ISHI-Veg) was established in 1994, herbage (ISHI-H) in 1997 and field crops (ISHI-F) in 1998, however, the latter two have not been active with only one method approved by ISHI-F (ISF 2013). ISHI-Veg started as an initiative of the vegetable seed industry when Dutch and French seed companies started a project on monitoring seed health in 1993. The USA, Japan and Israel joined the initiative which then represented more than 75 % of the world's

Fig. 2.2 Jim W. Sheppard and Valerie Cockerell

vegetable seed supply and in 2000 the International Seed Federation (ISF) took over the secretariat and financial administration (ISF 2013).

The Association of Official Seed Analysts (AOSA) formed in 1908, is an organization of member laboratories across the United States and Canada that establishes and publishes seed testing rules in the AOSA *Rules for Testing Seeds* (AOSA 2013). Although AOSA does have a seed pathology research sub-committee, it is the National Seed Health System (NSHS) that is involved in the evaluation of seed health methods. The NSHS, a program authorized by the United States Department of Agriculture-Animal and Plant Health Inspection Service (USDA-APHIS), was established in 1999 and published in the Federal Register on 18 July 2001 to come into effect on 17 August 2001 (Federal Register 2001). The NSHS is administered by the Iowa State University Seed Science Centre (NSHS 2013).

2 Seed Health Test Method Evaluation and Validation

2.1 ISTA

ISTA method validation critically examines a seed quality test to ensure that the description of the method is clear and complete and that the procedure gives accurate, reproducible and repeatable results (Hampton 2007). The operation of the formal ISTA Method Validation process relies on co-operation between the Technical Committee Coordinator (TCord) and the Technical Committee (TCOM) (in the case of seed health testing, the SHC). The TCord and the ISTA Secretariat support the TCOM by keeping records of the process and providing the TCOM chair with information they require to monitor the process. The validation process is described and published on the ISTA website as the *ISTA Method Validation for Seed Testing* (ISTA 2007) and the *Standard Operating Procedure for Method Validation Process Administration* (Hampton 2009; ISTA 2013) and is summarized below. ISTA also provides guidelines for ISTA rules proposals and instructions for authors and reviewers of draft test plans and validation reports on their website (ISTA 2013). Details and progress of method validation projects, which are the responsibility of the TCOM, are presented in an Excel workbook on the ISTA website.

ISTA seed health validated methods have been through multi-laboratory collaborative studies however, in some situations (for example: the addition of a new species to an existing method) ISTA has introduced its Peer-Validation Programme for the validation of methods by laboratories working with only one or two others (ISTA 2007). The validation process proceeds as follows (Fig. 2.3): If the SHC decides to proceed with a submitted test plan for a seed health test, two technical reviewers are appointed and the Statistics Committee (STA) appoints a statistical reviewer to review the test plan. The reviewed test plan is returned to the TCord.

The TCord forwards the reviews to the SHC who either rejects the test plan or approves it with or without minor revisions. The applicant of the approved test plan acts as or appoints a test organiser who drives the validation study, analyses of results and produces a validation report which is first sent to the validation study participants before being sent to the TCord. The TCord forwards the validation report to the SHC and STA who appoint two technical reviewers and one statistical reviewer, respectively. The reviewers return their reviews and the validation report to the TCord who forwards them to the SHC. The SHC either rejects the final validation report (an ISTA Method Validation Report) which includes all results, statistical data and a working method based on the reviews or approves them with or without minor revisions. The SHC prepares the method for the rules proposal on the basis of the validated report and the method becomes an ISTA Rules proposal (ISTA 2013). The method is published on the ISTA website to coincide with announcements of Rules proposals to be voted on at the next ISTA Ordinary Meeting providing the ISTA voting delegate with all the documentation to make an informed decision. The method requires the acceptance by the ISTA membership by vote at an Ordinary Meeting before publication of the validated method in the ISTA Rules. There are currently more than 29 validated ISTA seed health methods (some having been validated for the same pathogen but on different substrates) and are published as 7-001a to 7-029 on the ISTA website and in Chap. 7 of the ISTA rules (ISTA 2013).

2.2 *ISHI-Veg*

The method validation procedure followed by ISHI-Veg (ISF 2013) is summarized below and in Fig. 2.4. The International Technical Group (ITG) is responsible for developing a seed health test plan. There are four ITGs within the Technical Co-ordination Groups (TCG) namely, roots, bulbs and leafy vegetables; bean, pea and brassica; cucurbits; and tomato and pepper, that meet every 9–11 months. A nominated representative represents each country in the ITG but can consist of more than one representative per country depending on the crop/pathogen combination. Each TCG chairperson co-ordinates and maintains the ISHI-Veg comparative test plans developed by the ITGs. The TCG chairperson also represents the TCG in the ISF Phytosanitary Committee and the Policy Co-ordination Group (PCG), which consists of one member each from the participating countries. The PCG determines the annual budget of ISHI-Veg, develops policy guidelines to assist the TCG, decides which of the reference test methods developed by the TCG appear on the ISF website in the ISHI-Veg Manual and determines the recommended sample size for each test method. The PCG chair reports to both the ISF Vegetable Crop Section and the Phytosanitary Committee at the annual ISF Congress. Through the presentation of a draft motion and its subsequent adoption by the Vegetable Section a position taken by ISHI-Veg can become the position of ISF (Fig. 2.4). The seed health testing methods were the basis for the position

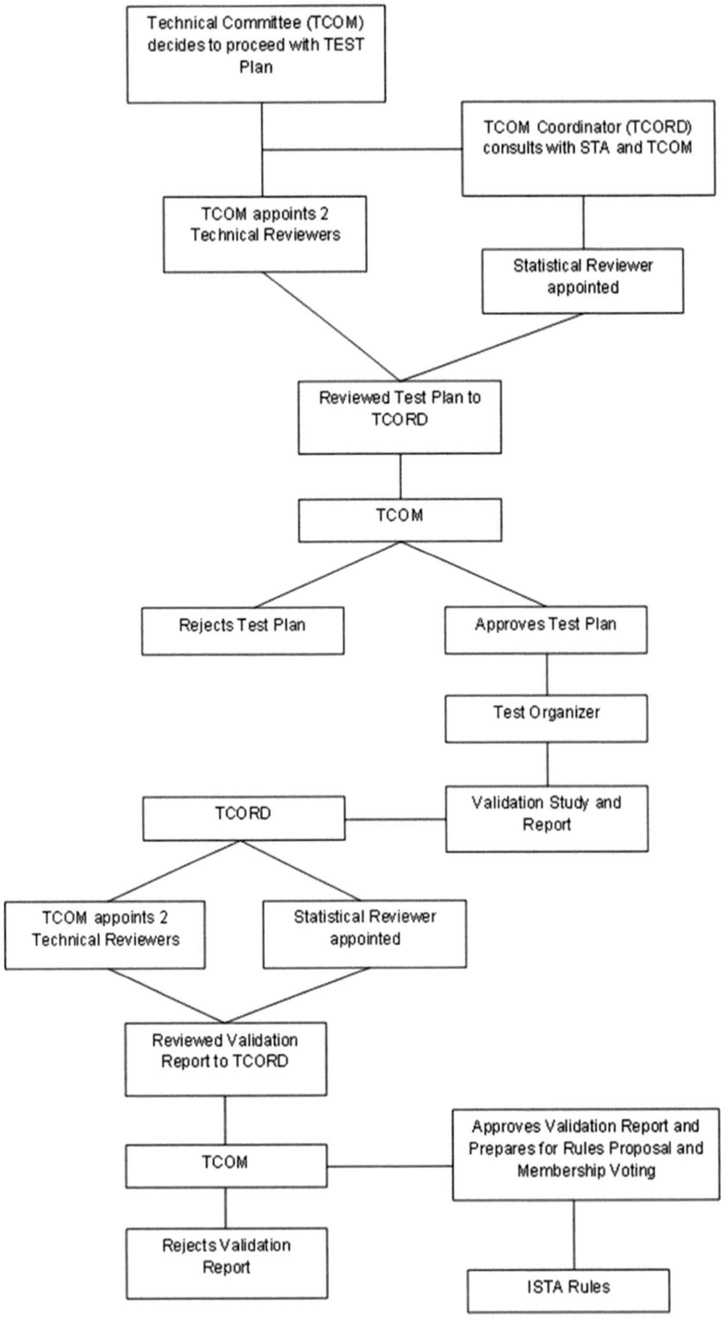

Fig. 2.3 The process followed by the Seed Health Committee of the International Seed Testing Association (ISTA) to validate a seed health test method

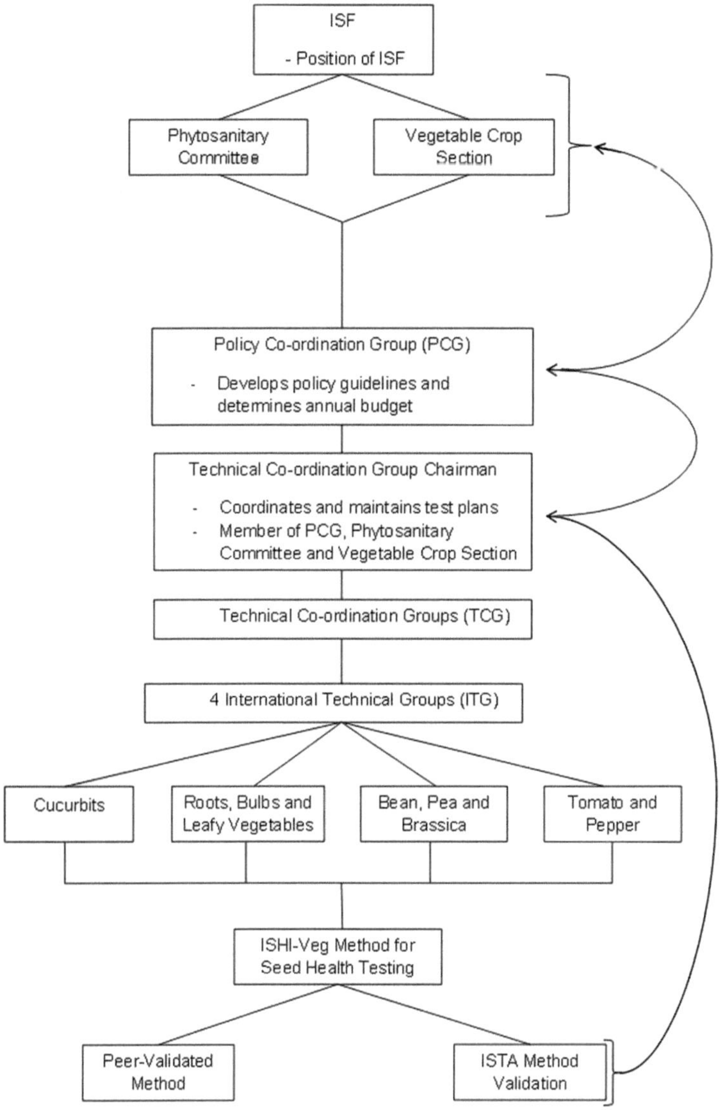

Fig. 2.4 The process followed by the International Seed Health Initiative for Vegetables (ISHI-Veg) of the International Seed Federation (ISF) to validate a seed health test method

adopted by the vegetable seed sector in May 2006 titled *Guidelines for the use of seed health methods by the vegetable seed industry*, which was revised in 2010.

The TCGs are composed of seed health scientists from public and private sectors and are responsible for the establishment of reliable test methods that (1) are clear and reproducible; (2) are practical and feasible for routine testing by technical staff; (3) give results that are indisputable; (4) function as generally accepted reference methods; (5) serve as legal references in court cases; (6) support the international seed industry in improving product quality and; (7) serve as documentation for phytosanitary certification according to IPPC guidelines (ISF 2013). For a method to be accepted by ISHI it must be described and available for public use or published in a peer-reviewed journal and must be approved by the TCG of the particular vegetable crop. The method then enters the validation process according to ISTA's Method Validation for Seed Testing (ISTA 2007) which includes a multi-laboratory comparative test of 6–8 company and public laboratories. The number of laboratories and samples are determined in the design of the test plan and if less than the six laboratories required by ISTA are used, the method may be accepted as a "Peer-Validated Method" and endorsed as an ISHI reference method. Data related to the validation of components of the method and the peer review are stored in the ISHI database (ISF 2013). Currently ISHI-Veg has 21 methods in their *Manual of Seed Health Testing Methods* of which 11 have been adopted as ISTA rules and 11 as NSHS standards (Munkvold 2009; ISF 2013).

2.3 NSHS

Reference Manual A. The reference manual for administration, procedures, and policies of the NSHS, published by the NSHS in 2000, describes the structure, administration, procedures, policies and working practices of the NSHS and also contains relevant documentation, forms and references for the NSHS (Federal Register 2001; NSHS 2013). The NSHS evaluation procedure of a new seed health testing method (NSHS 2013) is described below and in Fig. 2.5. The NSHS Administration Unit develops a technical package of scientific information on each host/pathogen combination. This includes published seed health tests, unpublished seed health tests in use, proprietary methodologies in use, data on test standardization by ISHI, ISTA or other organizations and relevant data on seed transmission or detection of the pathogen. This technical package is forwarded to a Technical Panel chosen from a number of seed health experts both nationally and internationally and from private and public sectors based on their area of expertise. Members of the Technical Panel volunteer to review existing and proposed new seed health testing methods and are approved by the Seed Technical Working Group (STWG). The Technical Panels use the following criteria to evaluate seed health test methods: (1) Empirical test data that determines the sensitivity, specificity, repeatability and reliability of the assay; (2) Comparative test data with already established methods; (3) Historical data of methods used in industry or academia and the record of number of uses or complaints and; (4) Other criteria which may have an impact on the recommendation for use of a test. Each Technical Panel reviewer prepares an individual report rating each of the methods evaluated by these criteria which is then submitted to the Technical Panel chair who prepares a summary technical report submitted to the STWG. The technical panel members use the following guidelines for rating the methodology(s):

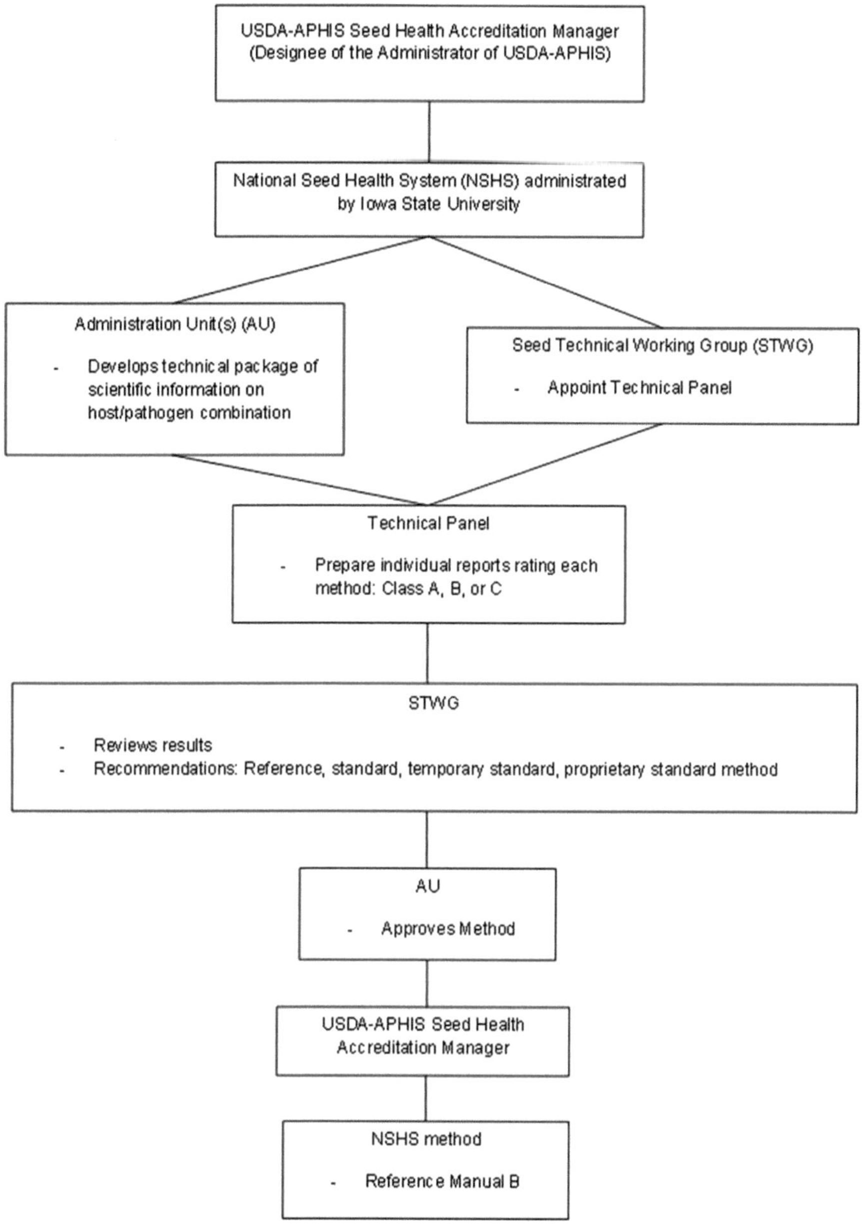

Fig. 2.5 The process followed by the United States of America National Seed Health System (NSHS) to evaluate a seed health test method

- Class A – The test or method acceptable as a standard test
- Class B – The test or method needs further research before acceptance as a standard test. This could be for improvement to the method itself or a recommendation for a comparative test with a known method
- Class C – The test should not be accepted as a standard test.

The STWG is responsible for reviewing the results and recommendations of the Technical Panel before forwarding them to the Administrator Unit for approval. There are four categories that the methods can be divided into:

1. Reference method – Recommended Class A and NSHS approved method accessible to all users
2. Standard method – Recommended Class A and NSHS approved method but method owner may recover royalties from use of the method
3. Temporary standard method – Recommended Class B and STWG approved for a defined period
4. Proprietary standard method – Comparable to an NSHS reference method and approved by the STWG.

All these methods may then be included in Reference Manual B for official use in phytosanitary certification following APHIS approval with exception of the proprietary standard method (NSHS 2013). *Reference Manual B. The reference manual for seed health testing and phytosanitary field inspection methods* published by the NSHS in 2001 contains the detailed seed health testing, seed sampling and inspection methods and procedures for the NSHS and is constantly updated (Federal Register 2001; NSHS 2013).

Conclusion

The procedures followed by the major role players in achieving global standards in seed health testing, namely ISTA, ISHI and the NSHS have been summarized. The procedures followed by the Seed Health Committee of ISTA and ISHI-Veg are very similar and are based on the validation procedure manual published by Sheppard and Cockerell (2000). ISTA and ISHI-Veg make use of collaborative multi-laboratory tests and elaborate statistical procedures and analyses to validate a new seed health test method. These procedures differ greatly from those followed by the NSHS. The NSHS appoints a panel that reviews scientific literature of existing seed health test methods to compare them with a new proposed method. There are no collaborative laboratory tests involved and hence, new methods are adopted and published by the NSHS in Reference Manual B far quicker than the adoption and publication of ISHI-Veg and ISTA methods. Methods adopted by ISHI-Veg and ISTA can be approved as NSHS methods but the reverse is not true unless collaborative multi-laboratory tests are conducted using the NSHS method. ISHI-Veg seed health methods often become ISTA methods.

(continued)

NSHS seed health methods are approved by the USDA-APHIS. In ISHI-Veg a new method is adopted by ISF, by both the ISF Vegetable Crop Section and the Phytosanitary Committee, after presentation at the annual ISF Congress. The new seed health method adoption process by ISTA involves several stages starting with the publication of all validation reports on the ISTA website and in *Seed Testing International* (ISTA 2013). The new method is sent out to the entire ISTA membership for perusal as part of the new methods to be adopted at the next ISTA ordinary meeting. At this meeting the new method is presented to the ISTA membership and can only be adopted as a new seed health method after a majority vote by each country's designated voting member. Furthermore, the seed health methods are reviewed every 5 years by the SHC and laboratories using the methods.

Although the global seed industry requires that the number of standardized seed health tests be exponentially increased, organizations involved with the standardization of new methods have to insure that the methods are validated and thus reproducible and repeatable by laboratories from all over the world. This is essential for the overall harmonization of phytosanitary regulations involving the movement of seed in international seed trade (Munkvold 2009).

References

Agarwal VK, Sinclair JB (1987) Principles in seed pathology, vol 1. CRC Press, Florida

Anon (1985) Fellows. P Paul, F M deNeergaard. Phytopathology 75(1):32–33

AOSA (2013) Association of Official Seed Analysts. http://www.aosaseed.com. Accessed 18 Aug 2013

Doyer LC (1938) Manual for the determination of seed-borne diseases. International Seed Testing Association, Wageningen

EPPO (1954) Danger from seed-borne diseases. A practical approach to the problem of international safeguard. Report on the working party on seed-borne diseases. EPPO, Paris

Federal Register (2001) Accreditation standards for laboratory seed health testing and seed crop phytosanitary inspection. Fed Regist 66(138):37397–37401

Hampton J (2005) ISTA method validation. Seed Test Int 130:22–23

Hampton J (2007) ISTA method validation. Seed Test Int 133:39

Hampton J (2009) Report of the ISTA method validation advisory group. In: ISTA annual meeting 2009, Zurich

ISF (2013) The International Seed Health Initiative (ISHI). http://www.worldseed.org/isf/ishi.html. Accessed 14 Aug 2013

ISTA (2007) ISTA method validation for seed testing. International Seed Testing Association, Basserdorf

ISTA (2013) The International Seed Testing Association (ISTA). http://www.seedtest.org. Accessed 31 Jul 2013

Mathur SB, Jørgensen J (2002) A review of the activities of the plant disease committee of ISTA through its 75 years of existence, 1924–1999. ISTA Hist Pap 1:1–34

Munkvold GP (2009) Seed pathology progress in academia and industry. Annu Rev Phytopathol 47:285–311

Muschick M (2010) The evolution of seed testing. Seed Test Int 139:3–7

Neergaard P (1970) Seed pathology, international co-operation and organization. In: Proceedings of the International Seed Testing Association, vol 35, Copenhagen, pp 19–42
NSHS (2013) National seed health system. http://www.seedhealth.org/. Assessed 14 Aug 2013
Sheppard J, Cockerell V (2000) ISTA handbook of method validation for the detection of seedborne pathogens, Plant disease communication. International Seed Testing Association, Zurich
Sheppard JW, Wesseling JBM (1998) ISTA/ISHI guide for comparative testing of methods for the detection of seed-borne pathogens. Seed Sci Technol 26:237–255
Steiner AM, Kruse M, Leist N (2008) ISTA method validation 2007: a historical retrospect. Seed Test Int 136:30–33

Chapter 3
Seed-Borne Pests and Phytosanitary Issues: the Role of EPPO

Françoise Petter, Muriel Suffert, Madeleine McMullen, Daminen Griessinger, and Anne-Sophie Roy

Abstract Since a number of important pests are seed-borne or seed-transmissible, the movement of infested seed may pose a risk for the international spread of pests. One of the main roles of the European and Mediterranean Plant Protection Organization (EPPO) is to help its member countries to prevent entry or spread of dangerous pests. The Organization has therefore been given the task of identifying pests which may present a risk for the region (early warning), evaluating them and making proposals on the phytosanitary measures which can mitigate the risk. Once a pest has been identified as presenting a risk for the EPPO region, guidance on how to detect and identify it are developed (phytosanitary procedures for inspection and diagnostic protocols) as well as recommendations on how to eradicate and/or control it. To perform these activities, EPPO collects information and makes it available to its member countries. Different databases have been developed including PQR (Plant Quarantine data Retrieval system) and the EPPO database on Diagnostic expertise. In addition to pest-specific activities, EPPO also develops recommendations for quality assurance in laboratories, in order to harmonize procedures in the EPPO region and improve diagnostic quality. The different activities conducted in this framework are presented with a special focus on activities related to seed-borne pests.

Keywords Pest risk analysis • Diagnostics • Plant quarantine • EPPO

1 Background

Ensuring health and quality of seeds is the first step in the production of food and feed. Nowadays, many seeds are moved internationally, primarily for food and ornamental plant production, but also for example, for production of biofuels, fibre, pharmaceuticals as well as for pre-commercial (research, seed increase) purposes. Therefore mechanisms are needed that ensure the safe movement of seeds in international trade in order to protect agriculture and the environment. For this

F. Petter (✉) • M. Suffert • M. McMullen • D. Griessinger • A.-S. Roy
European and Mediterranean Plant Protection Organization, 21 Boulevard Richard Lenoir, 75016 Paris, France
e-mail: petter@eppo.int

M.L. Gullino, G. Munkvold (eds.), *Global Perspectives on the Health of Seeds and Plant Propagation Material*, Plant Pathology in the 21st Century 6,
DOI 10.1007/978-94-017-9389-6_3

reason, most countries have phytosanitary requirements on the movement of seed, though there is significant variation in the methods used for assessing the phytosanitary risk associated with seed, the pests of concern, phytosanitary import requirements, diagnostic and inspection methodologies, and acceptable phytosanitary risk mitigation measures.

The article explains the role of EPPO in this framework.

1.1 International context of Plant Health and the birth of Regional Plant Protection Organizations (RPPOs)

Human societies have throughout their histories faced the emergence of pests which damaged crops or the environment. In plant pathology, the classical example remains the disastrous consequences of the introduction of potato late blight (*Phytophthora infestans*) which caused famine in Ireland in the 1840s and now causes problems in potato production worldwide. The need to prevent the movement of dangerous pests when moving plants between countries has consequently been recognized in the late nineteenth century with the adoption of international conventions such as the International Convention respecting measures to be taken against *Phylloxera vastatrix* in 1881 (following the introduction of grapevine phylloxera in Europe) the additional Convention signed at Berne on 15 April 1889 and the International Convention for the Protection of Plants in 1929. After the Second World War a more permanent collaboration developed and these conventions were replaced in 1951 by the International Plant Protection Convention (IPPC) revised in 1979 and 1997. The IPPC is an international plant health agreement that aims to protect cultivated and wild plants by preventing the introduction and spread of pests. The definition of pest as given in the Convention is '*any species, strain or biotype of plant, animal or pathogenic agent injurious to plants or plant products*' (i.e. bacteria, fungi, insects, plants, viruses…). The IPPC is recognized by the World Trade Organization agreement as the relevant international standard setting organization for plant health matters. In 2013, 179 countries were contracting parties to the Convention. The implementation of the IPPC involves the collaboration of National Plant Protection Organizations (official services established by governments to carry out the functions specified by the IPPC), but the IPPC also includes provisions for the establishment of Regional Plant Protection Organizations (RPPOs), which function as coordinating bodies at a regional level for activities to achieve the objectives of the Convention. Since 1951, 10 RPPOs have been created among which EPPO is the oldest. RPPOs are intergovernmental organizations (i.e. their official members are countries not individuals); most of them have been founded on the initiative of governments, while others are administered by FAO regional offices. The existing RPPOs are listed in Table 3.1.

Most countries worldwide are included in one or more RPPO. It can be noted that, due to their geographical position, some countries are entitled to join more

Table 3.1 List of Regional Plant Protection Organizations in 2014

Asia and Pacific Plant Protection Commission (APPPC), with 24 member countries
Caribbean Plant Protection Commission (CPPC), with 22 member countries
Comité Regional de Sanidad Vegetal Para el Cono Sur (COSAVE), with 7 member countries
Comunidad Andina (CA), with 4 member countries
European and Mediterranean Plant Protection Organization (EPPO), with 50 member countries
Inter-African Phytosanitary Council (IAPSC), with 54 member countries
Near East Plant Protection Organization (NEPPO) with 10 member countries
North American Plant Protection Organization (NAPPO), with 3 member countries
Organismo Internacional Regional de Sanidad Agropecuaria (OIRSA), with 8 member countries
Pacific Plant Protection Organization (PPPO), with 21 member countries

than one organization (e.g. Algeria, Morocco, Jordan and Tunisia are member of both EPPO and NEPPO). In recent years, the RPPOs started to cooperate more closely and they now meet once a year for a Technical Consultation. RPPOs also actively participate with their member countries in the preparation of International Standards on Phytosanitary Measures (ISPMs).

1.2 The European and Mediterranean Plant Protection Organization

EPPO was created in 1951, the same year as the adoption of the IPPC. Initially founded by 15 member governments it now has 50 member countries. It includes nearly all Western and Eastern European countries, Mediterranean basin countries as well as countries from Central Asia (see Fig. 3.1).

The aims of EPPO as given in the Convention (EPPO, last revised 1999) are:

- To protect plant health in agriculture, forestry and the uncultivated environment.
- To develop an international strategy against the introduction and spread of pests (including invasive alien plants) that damage cultivated and wild plants, in natural and agricultural ecosystems.
- To encourage harmonization of phytosanitary regulations and all other areas of official plant protection action.
- To promote the use of modern, safe, and effective pest control methods.
- To provide a documentation service on plant protection.

Recommendations of EPPO are mainly issued in the form of Standards and EPPO's programme of activity is directed by two Working Parties (on Phytosanitary Regulations and on Plant Protection Products). Specific tasks are assigned to Panels composed of specialists from EPPO member countries. The technical work of the Organization consequently depends on the active and

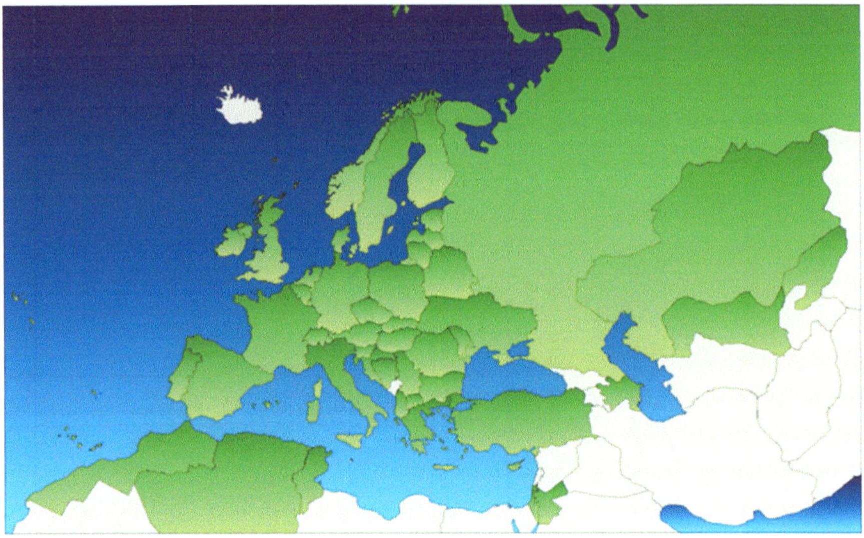

Fig. 3.1 EPPO member countries (in *green* online/*grey* in print)

continued participation of experts. Recommendations of EPPO are used by EPPO members to prepare their phytosanitary regulations.

With regards to the aim to prevent entry or spread of dangerous pests (plant quarantine) the programme of activity established is presented in this article with examples given for seed-borne pests whenever relevant.

Seed-Borne Pests: some Terminology

Difference between seed and grain

Seed is defined as "A commodity class for seeds for planting or intended for planting and not for consumption or processing" (ISPM 5).

Grain is defined as "A commodity class for seeds intended for processing or consumption but not for planting (see seeds)" (ISPM 5).

From a phytosanitary point of view, they are quite different as the risk to spread pests associated with seed is higher than with grain as seed will be planted.

Seed-borne and seed-transmitted pests

Seed-borne pest are those that can be found on the seed or within the seed coat but do not necessarily result in the transfer of the pests to the resulting plant. Seed transmitted pests are those that can be transferred from the seed into the resulting plant.

During the PRA process, care should be taken to ensure that seed is actually a pathway for the introduction of any regulated pests for which phytosanitary measures are to be required.

2 Current EPPO activities in Plant Quarantine

One of the consequences of the increase in international trade in recent years is that countries have been faced with the introduction of several new pests. Consequently, more than ever, regional collaboration is needed to collectively face this challenge and RPPOs have a clear role to play. EPPO adopted a strategy for 2010–2014 which is taking into account this international context to better support the work of its member countries.

To prevent entry or spread of dangerous pests within the EPPO region, the Organization has been given the task of identifying pests which should be recommended for regulation by its members. Consequently since the 1970s, EPPO has maintained a List of A1 and A2 pests recommended for regulation which currently contain more than 300 pests. The list distinguishes pests which are absent (A1) from the EPPO region from those which are locally present (A2). The World Trade Organization Sanitary and Phytosanitary Agreement requires that phytosanitary measures must be justified and commensurate to the risk, and measures adopted by countries to protect their territories from these pests should be technically justified. Therefore EPPO follows a highly transparent process prior to listing which consists of:

- identifying pests which may present a risk (early warning),
- evaluating their risk for the region and making proposals on the phytosanitary measures which can be taken against them, a process called Pest Risk Analysis (PRA).

Once a pest has been identified as presenting a risk for the EPPO region, recommendations on how to detect and identify the pest are developed (diagnostic protocols and phytosanitary procedures for inspection) as well as recommendations on how to eradicate and control this pest. To perform these activities, EPPO collect information and makes it available to its member countries. Details on these activities are provided below.

2.1 *Activities on Early Warning*

When pests are emerging in the EPPO region or in other parts of the world, it is necessary to provide early warning to NPPOs so that they can put into place import inspections or surveillance programmes in their territory. Since 1998, EPPO has established an Alert List to provide data on emerging pests (the current list includes the following seed-borne pests *Acidovorax citrulli*, *Fusarium oxysporum* f.sp. *lactucae*, *Tomato apical stunt pospiviroid*). The current content of the EPPO Alert List can be viewed on the EPPO website: http://www.eppo.int/QUARANTINE/Alert_List/alert_list.htm. Some of these emerging pests may later be submitted to a PRA and depending on the outcome of the assessment they may be recommended for regulation as quarantine pests. This was the case with *Xanthomonas axonopodis* pv. *allii* a pest recommended for regulation in 2009 and *Sirococcus clavigignenti-juglandacearum* in 2005.

Pests on the Alert List are often selected by the EPPO Secretariat, mainly from the literature and internet surveys, but they are increasingly included following the request of Plant Protection Services or individual experts from EPPO member countries. All pests on the Alert List are selected because they may present a phytosanitary risk for the EPPO region, but are not necessarily of concern to other parts of the world. The reasons for considering inclusion on the Alert List include: pests which are new to science, new outbreaks in the EPPO region or elsewhere, reports of spread in the EPPO region or elsewhere, new host plants etc. For each pest, a mini datasheet is provided and includes a brief summary on the reasons why the pest was added to the EPPO Alert List, its geographical distribution, main host plant, type of damage, mode of dissemination, possible pathways of introduction and preliminary elements of risk. It is important to note that the EPPO Alert List is not a quarantine list, and does not constitute a recommendation for phytosanitary action. The section 'possible risk' is not the result of a PRA according to EPPO Standard PM 5/3(1) but is a preliminary attempt by the EPPO Secretariat to identify the main elements of risk. The Alert List is reviewed critically every year by the EPPO Panel on Phytosanitary Measures and some of the pests may later be selected and submitted to a PRA. As already stated, pests may be added to the EPPO A1 and A2 Lists of pests recommended for regulation or, if the PRA shows the risk to be low, they can be removed from the Alert List. This was the case for *Claviceps africana* based on an Italian PRA concluding that the risk was low for the EPPO region and for *Puccinia psidii* where a French PRA concluded that the risk was low due to the climatic requirements of this rust. For practical reasons, each pest entry is kept on the Alert List for 3 years, after this period the pest can be deleted from the Alert List if no particular interest was shown by the EPPO member countries. The information on the organisms which were previously listed on the EPPO Alert List remains available on the EPPO website and the information which was available at the time when they were deleted can be retrieved there (http://www.eppo.int/QUARANTINE/Alert_List/deletions.htm).

2.2 Pest Risk Analysis activities

Pest Risk Analysis (PRA) as defined by the IPPC is: "*The process of evaluating biological or other scientific and economic evidence to determine whether an organism is a pest, whether it should be regulated, and the strength of any phytosanitary measures to be taken against it*". Since the early 1990s, EPPO has established a work programme on PRA which first consisted in preparing regional Standards for PRAs:

- PM 5/1 *Check-list of information required for pest risk analysis (PRA)* approved in 1992,
- PM 5/2 *Pest risk analysis on detection of a pest in an imported consignment*, approved in 1992 and revised in 2001

While contributing to the development of the ISPMs on Pest Risk Analysis such as ISPM no. 11 "*Pest risk analysis for quarantine pests*" (IPPC 2013), EPPO has also developed a regional scheme for PRA now called the *EPPO Decision-support scheme for pest risk analysis of quarantine pests* (EPPO 2011). This scheme has the added value of guiding the assessor through a logical sequence of questions covering all elements mentioned in ISPM no. 11. The scheme was revised in 2011 in the framework of the European Union 7th framework program protect PRATIQUE (Baker 2012). A piece of computer software named CAPRA has also been developed by the EPPO Secretariat in the framework of PRATIQUE and with the support of the EPPO Panels (Griessinger et al. 2012). This computer system aims to assist pest risk analysts to run the EPPO Decision-Support scheme for PRA and is freely available on the EPPO website (http://capra.eppo.int). EPPO has also recently developed a *Decision-Support Scheme for an Express Pest Risk Analysis* which provides a simplified scheme for the rapid production of pest risk analyses (EPPO 2012).

In order to add pests to the EPPO List of pests recommended for regulation, emerging pests identified during early warning activities of EPPO and its member countries are submitted to PRAs. EPPO has been given the task by its members to perform PRAs on a regional scale.

As already explained in Petter et al. (2010) PRAs prepared in the EPPO framework are usually conducted following the *EPPO Decision-support scheme for PRA* however since 2012, they may also be prepared following the *Decision-Support Scheme for an Express Pest Risk Analysis* (a recommendation on the scheme that should preferably be used is made by the EPPO Panel on Phytosanitary Measures). Depending on the pest to be evaluated, a combination of both schemes may also be used with some parts of the assessment following the Express PRA scheme (e.g. when it is clear that establishment is very likely to happen and does not need a detailed evaluation because the pest is has established in one part of the region) whereas other are answered in more details following the Decision-support scheme for PRA (e.g. economic impact, management part...). The output of a PRA takes the form of a general recommendation to countries, with measures proposed for each organism concerned, distinguishing different levels of risk for different parts of the EPPO region as applicable. This recommendation has then to be adopted by consensus by the EPPO Members, after appropriate consultation. Members decide individually whether the reported risks concern them, and select appropriate measures if they do. The EPPO Convention creates no greater obligation on members than that they should "endeavour to implement" EPPO recommendations. However, there is a general policy of "regional solidarity", by which Members take phytosanitary measures against pests which are not present in the

EPPO region and select their measures from those recommended. Countries may choose not to apply these measures if the risk of establishment on their territory is very low, e.g. if climatic requirements are not met. The PRA documents reviewed and elaborated in the EPPO framework (PRA records, PRA reports, datasheets) are freely available on the EPPO website (www.eppo.int).

2.2.1 Since 2006, a new system established for the preparation of PRA at the EPPO level

In 2004 and 2005, the role of EPPO in PRA was discussed both at the political level and at the technical level. It was recognized that many countries do not have the resources to perform PRA and consequently member countries wished that EPPO should play an active role in organizing internationally conducted PRA in the region, in order to share costs and workload and to provide technical justification for the regulation of certain pests. The proposal that special EPPO Expert Working Groups for PRA (called later PRA EWG) should now perform PRAs, as well as reviewing PRAs from other sources was made. It was also considered that the creation of a specialized PRA EWG would encourage collaboration between members and increase the quality of the PRAs produced. On average five PRA EWG are organized per year on different pests. Unlike other groups in EPPO, the PRA EWGs have a varying composition and experts on specific pests can be called upon to participate when needed. The PRA EWG also have core-members to provide consistency. These core-members are usually drawn from existing EPPO Panels with experience in performing or reviewing risk assessment and determining risk management options such as the Panel on Phytosanitary Measures and the Panel on Quarantine Pests for Forestry. For each PRA EWG the EPPO Secretariat tries to balance the experience within the group. Groups include experts on the pest or group of pests to be studied, experts in risk management, experts on the crop concerned, experts with knowledge in running the EPPO PRA schemes, experts on tools to help predict the future distribution of the pest (e.g. GIS, Climex) and whenever possible an expert in socio-economics, although this latter objective has proved difficult to achieve. When the pest evaluated is absent from the EPPO region, it is common that experts from the area of origin of the pest are invited so that the group can benefit from their practical experience with the pest. Experts help the EPPO Secretariat to gather of necessary information to prepare the PRA. Recently a consultant has been contracted to prepare a pre-PRA so that essential missing information is identified before the meeting of the PRA EWG takes place.

The pre-PRA is reviewed in detail during the meeting of the Expert Working Group, following the relevant (or combined) EPPO Decision-support scheme. The Expert Working Group goes through each individual question on the scheme. Each answer should be justified and justifications are recorded in a document named PRA record which is prepared during the meeting. A datasheet on the pest should also be prepared, preferably by one of the experts on the pest.

The PRAs, are sent by email to a group of reviewers who are the core members. When comments are made, the PRA EWG is consulted by email. The PRAs are subsequently presented to the relevant bodies which decide by consensus on the appropriate recommendation to be made to EPPO member countries.

Example of a PRA conducted for a Seed-Borne Pest

Xanthonomas axonopodis pv. *allii* (a bacterium of *Allium* spp.) was identified as an emerging pest and added to the Alert List. A PRA EWG was convened and two experts from areas where this bacterium is present were invited to participate, Mr Gent USDA-ARS, Forage Seed and Cereal Research Unit, United States and Mr Pruvost CIRAD UMR PVBMT Reunion Island, France. Some specific elements of the PRA are highlighted below. The PRA is available on the EPPO website in the section PRAs conducted by EPPO Expert Working Groups http://www.eppo.int/QUARANTINE/Pest_Risk_Analysis/PRA_intro.htm.

With a seed-borne pest, an important factor to be considered is not only the association of the pest with seeds but the fact that it is transmitted by seeds to the progeny. This is essential for the transfer of the pest to occur and this is a necessary step for a pest to be able to establish in a new area. In the case of *Xanthonomas axonopodis* pv. *allii* seed transmission of the disease had been demonstrated and was suspected to be the pathway for the introduction of the pathogen to Reunion Island.

Concentration of the pest on this pathway is another element considered during the evaluation: one should consider whether practices, mainly in the country of origin, such as cultural practices (e.g. plant protection products, growing conditions) will decrease the concentration of the pest on the pathway. For example it was considered that for *X. axonopodis* pv. *allii* management practices lead to disease suppression but that seed contaminated at a rate of 4 seeds per 10,000 were able to induce an outbreak (Roumagnac et al. 2004). It was also noted that no seed treatment is available.

Regarding the probability of establishment it was noted that *Allium* spp. are widely grown in the EPPO region but that given the climatic requirement of *X. axonopodis* pv. *allii*, Mediterranean countries were considered more at risk than temperate countries, and northern countries were not considered at risk. Some cultural practices favour the infection of crops and spread of the pest including overhead irrigation which is common in several EPPO countries.

During the assessment of potential economic consequences it was noted that the bacterium can cause significant yield losses and high control costs when conditions are suitable. It negatively affects bulb size because of the destruction of the foliage. In the continental United States, yield losses ranging from 10 % to 50 % are reported. The negative effect was

(continued)

consequently considered as major. It should be noted that information on the economic impact of a pest is not always easy to find.

The risk was considered as not acceptable and management options were identified for seeds of *Allium* spp. which are as follows:

- Seeds should be produced in pest-free areas or pest-free places of production

A place of production freedom should consist of a combination of the following individual measures:

- The pest should have been absent from the place of production in the previous growing period (based upon inspection and testing)
- Sanitation measures in the growing crop (e.g. prevention of infection with tools, equipment, etc.)
- Seeds produced from seeds (or bulbs) which are free from the pest.
- Buffer zone of 1 km to 5 km depending on local climatic conditions (e.g. in areas prone to storms). *It was noted that there is uncertainty on the minimum distance needed for the buffer zone.*
- Testing during the growing period.

It was also recommended that importing countries may consider including *X. axonopodis* pv. *allii* in their surveillance programme and prepare an contingency plan for its eradication.

***X. axonopodis* pv. *allii* was added to the A1 List of pests recommended for regulation in 2009.**

PRAs conducted by EPPO member countries or other bodies such as the European Food Safety Authority (EFSA) are also used in the EPPO framework to make recommendations to EPPO members. An assessment of these PRAs is made to evaluate if the conclusions are also relevant for the EPPO region. A PRA on *Acidovorax citrulli* was conducted in the framework of an EFSA project Prima Phacie (MacLeod et al. 2012) this PRA was sent to the core-members and reviewed at the meeting of the Panel on Phytosanitary Measures in 2013. A possible recommendation for addition to the list of pests recommended for regulation is to be presented to the EPPO Council in September 2014 (for updates please see the EPPO website).

The EFSA Plant Health Unit and the EPPO Secretariat have established regular contact to share their respective work plans and avoid duplication of work.

The process described here has been formalized rather recently as the first EPPO activities on PRA were initiated in the late 1990s. In the past, the addition of pests to the EPPO List of pests recommended for regulation was based on an evaluation of

Table 3.2 Examples of seed-borne pests recommended for regulation by EPPO and related EPPO Standards

Pest	Type of pest	A1/A2	Year of listing	EPPO Standard
Clavibacter michiganensis subsp. *insidiosus*	Bacterium	A2	1975	PM 7
Clavibacter michiganensis subsp. *michiganensis*	Bacterium	A2	1975	PM 7; PM 3 in preparation
Curtobacterium flaccumfaciens pv. *flaccumfaciens*	Bacterium	A2	1975	PM 7
Gibberella circinata	Fungus	A2	2002	PM 7
Glomerella gossypii	Fungus	A2	1975	
Mycosphaerella dearnessii	Fungus	A2	1975	PM 7
Pantoea stewartii	Bacterium	A2	1975	PM 7
Pepino mosaic virus	Virus	A2	2012	PM 7
Stenocarpella macrospora	Fungus	A2	1975	
Stenocarpella maydis	Fungus	A2	1975	
Tilletia indica	Fungus	A1	1975	PM 7, PM 3
Xanthomonas axonopodis pv. *allii*	Bacterium	A1	2009	PM 7 in preparation
Xanthomonas axonopodis pv. *phaseoli*	Bacterium	A2	1975	
Xanthomonas oryzae pv. *oryzae* and *oryzicola*	Bacterium	A1	1975	PM 7
Xanthomonas translucens pv. *translucens*	Bacterium	A2	1993	
Xanthomonas spp. causing bacterial spot of tomato and sweet pepper	Bacterium	A2	1984	PM 7
…. and potato diseases (seed-tubers)	Various	A1 and A2		PM 7, PM 3, PM 9

PM 3, Phytosanitary procedures; PM 7, diagnostic protocol; PM 9, National regulatory control systems

technical information provided by experts and peer reviewed by the EPPO Panel on Phytosanitary Measures. Elements of justification of these additions can be found in the EPPO Datasheets (all freely available from the EPPO website).

Examples of seed-borne pests recommended for regulation by EPPO are given in Table 3.2.

2.2.2 Pathway Analysis

More recently, member countries have encouraged EPPO to start performing reviews of pests potentially associated with specific pathways, in addition to pest specific PRA. The first study concerned the tomato fruit pathway. The process consisted of identifying the pests that are likely to be associated with a pathway and establish prioritized lists of pests for which a pest specific PRA may be carried out, information on the pathway itself is also gathered (i.e. production modes, transport conditions, handling packing. . .). This would complement the early warning system by following a more systematic approach for a specific pathway. At present, there are no plans in the short term to carry out a pathway analysis on a seed commodity. However, the CPM plans to develop an ISPM on International movement of seed (see Sect. 3.2.6).

When a pest is recommended for regulation, EPPO supports its member countries by providing information on the pests and, for some of the pests, by developing specific Standards. Priorities for the development of specific Standards are set by EPPO bodies where member countries are represented.

2.3 Development of Standards in the Area of Diagnostics

In 1998, a programme was initiated to develop diagnostic protocols for as many as possible of the pests of the EPPO A1 and A2 lists (Zlof et al. 2000; Petter et al. 2013). The preparation of protocols involves close collaboration between different Panels composed of diagnostic experts:

- Panel on Diagnostics and Quality Assurance
- Panel on Diagnostics in Bacteriology
- Panel on Diagnostics in Entomology
- Panel on Diagnostics in Nematology
- Panel on Diagnostics in Virology and Phytoplasmology
- Panel on Diagnostics in Mycology

In total, these Panels involve about 100 experts.

Each draft diagnostic protocol is initially prepared by an individual expert according to a common format which ensures that the draft contains all necessary information to detect and positively identify a particular pest. Whenever available, validation data is also provided for the different tests included in the diagnostic protocols. The draft protocols are reviewed by the relevant Panels and submitted to a consultation phase among all EPPO member countries to ensure their wide acceptance. As it is the case for all EPPO Standards, diagnostic protocols are officially approved by the EPPO Council (at its yearly Session in September) and then published in the EPPO Bulletin and on the EPPO website. They are also freely available from the Phytosanitary Resources website hosted by the IPPC. In 2013, 119 diagnostic protocols have been approved.

EPPO is also active in the development of International Standards for Phytosanitary Measures for internationally agreed diagnostic protocols (ISPM 27 and its annexes).

Other regional plant protection organizations (e.g. NAPPO; www.nappo.org) are developing diagnostic protocols as well as individual countries (e.g. Australia;

http://www.padil.gov.au/). Such protocols are taken into account when developing EPPO protocols.

2.3.1 Projects and Input into EPPO Standard Development

The EPPO Secretariat works closely with other organizations and with consortia which are undertaking projects relevant to diagnostics. For example during the development of the recently published EPPO diagnostic protocol on *Pepino Mosaic Virus* the results from the EU FP6 PEPEIRA project were taken into account when selecting tests to be included in the protocol. EPPO was also closely associated with the EU Framework Research project QBOL on *Development of a new diagnostic tool using DNA barcoding to identify quarantine organisms in support of plant health*. EPPO, the QBOL partners, and the Dutch Plant Protection service, organized a joint Conference on DNA Barcoding and diagnostic methods for plant pests. One of the outcomes of this project will be an EPPO Standard on *DNA barcoding as identification tool for some regulated plant pests* which is under development. EPPO is currently a partner in the EU FP7 TESTA project *Seed health: development of seed treatment methods, evidence for seed transmission and assessment of seed health*. The Secretariat will work closely with the partners developing and validating new and existing diagnostics tests for seeds and will organize expert working groups so that at the end of the project EPPO protocols will be ready for adoption and publication. Priority will be given to protocols on *Ditylenchus gigas* and *Ditylenchus dipsaci* (a revision of an existing EPPO protocol and addition of *Ditylenchus gigas*), *Acidovorax citrulli* (new protocol), *Clavibacter michiganensis* subsp. *michiganensis* (revision of an existing EPPO protocol) in addition work on other regulated pests (*Tilletia* spp., *Pantoea stewartii*, *Pepino mosaic virus* and Pospiviroids) will also be used to prepare and revise EPPO Diagnostic protocols.

2.3.2 Accreditation and Quality Management

Two Standards on quality assurance have been developed so far PM 7/84 *Basic requirements for quality management in plant pest diagnosis laboratories* (EPPO 2007) and PM 7/98 *Specific requirements for laboratories preparing accreditation for a plant pest diagnostic activity* (EPPO 2010). A joint communiqué between EPPO and EA (European Co-operation for Accreditation, the European network of nationally recognised accreditation bodies) states that "*EA will recommend that assessors from Accreditation Bodies take note of EPPO documents when evaluating plant pest diagnostic laboratories*". EPPO also organized two workshops on quality assurance in 2007 and 2009, to allow experts to share their experience on quality assurance and accreditation. Workshops for Heads of Laboratories were held in 2011 and 2013 and topics included experiences with accreditation, organization of proficiency testing and test performance studies and scope of regional and national reference laboratories. A further Workshop on accreditation and quality assurance for laboratories was held in 2014.

2.3.3 Widely Used Tests

A survey on the use of the diagnostic protocols was conducted in 2008 on a selection of 58 protocols in all disciplines of plant health diagnosis (Petter and Suffert 2010). Laboratories registered in the EPPO database on Diagnostic Expertise (see below) were asked to indicate the number of samples that they tested in 2007 and which test they used. From this survey it could be concluded that many of the tests for detection mentioned in EPPO diagnostic protocols are widely used in laboratories in the EPPO Region. This survey to elaborate the list of tests widely used in the EPPO region was being repeated for tests carried out in 2012. The results from this survey are included in the Appendix of the Standard PM 7/98 (2) *Specific requirements for laboratories preparing accreditation for a plant pest diagnostic activity.*

2.3.4 EPPO Database on Diagnostic Expertise

In 2004, EPPO Council stressed that the implementation of phytosanitary regulations for quarantine pests was jeopardized by decreasing knowledge in plant protection. The Panel on Diagnostics proposed that an inventory should be made of the available expertise on diagnostics in Europe. The database on Diagnostic Expertise was created (Roy et al. 2010) to allow identification of experts who can provide diagnosis of regulated species and those who can help in the identification of new or unusual species. EPPO member countries were contacted and as of August 2013 more than 100 official diagnostic laboratories of the EPPO region have provided details about the pests they can diagnose and the methods they use corresponding to more than 500 experts). These results are available in a searchable database on the EPPO website. The database can also help national accreditation bodies identify technical auditors or technical experts for pest diagnostic laboratories for accreditation. A new section 'validation data for diagnostic tests' was added to the database on EPPO diagnostic expertise in December 2012 to share validation data generated by registered laboratories in the EPPO region.

The EPPO Secretariat considers that these initiatives and future plans will aid the optimization of diagnostic activities in laboratories in the EPPO region.

2.4 *Development of Standards on Inspection and Official Control*

2.4.1 Standards on Inspection

EPPO has initiated the development of Standards providing guidance on methods to be followed for performing inspections of commodities moving in trade, or surveys of quarantine pests. The first Standards were adopted in the 1980s, however, this activity has been dormant for a number of years and was only reactivated in 2013.

Most Standards in this series are in need of revision; Standards on the inspection of consignments of tomato seeds and of consignment of wheat grain and seed are in preparation and are expected to be ready for member consultation in 2014.

Procedures for consignment inspection include:

- a description of the commodities concerned and a section on the risk associated with the different commodities.
- Information on the pests of concern for the commodity (specific pests, polyphagous pests and contaminating pests) and elements of detection.
- The main means of lot identification
- General guidance on sampling for visual inspection and sampling for testing in the laboratory (including recommendations on minimum sampling level).

2.4.2 Standards on Official Control

Because of the recent increase in pest introductions NPPOs face the challenge of how to respond rapidly and effectively to pest outbreaks. EPPO members are consequently aiming to develop contingency plans for pests which may cause a major economic and/or environmental impact. In order to support its members EPPO developed Standard PM 9/10 *Generic elements for contingency plans* (EPPO 2009). No specific PM 9 has been developed for seed-borne pests but several have been established for potato-tuber transmitted pests. A decision support system for the eradication and containment of pest outbreaks is also in the final approval stages (it started in the framework of the EU project PRATIQUE).

2.5 *Provision of Information to EPPO Member Countries*

In its own Convention, EPPO has a clear task dedicated to information exchange. Each member country has to report on the existence, outbreak or spread of pests to EPPO, which in turn has to convey this information to all its members. Since its creation, EPPO has provided a Reporting Service to its member countries. In its present form, the EPPO Reporting Service is a monthly newsletter which reports on events of phytosanitary concern and focuses on new geographical records, new host plants, new invasive species (pests and diseases as well as invasive alien plants). This newsletter contains official reports made by NPPOs as well as information which is collected by the EPPO Secretariat from the scientific literature or other sources (see Fig. 3.2). The EPPO Reporting Service can be obtained freely by e-mail by any interested person. Information which is collected (e.g. on geographical distributions and host plant lists of many pests, including invasive species) is then stored in a database (PQR). Most of this pest-specific information is freely accesible from the EPPO website.

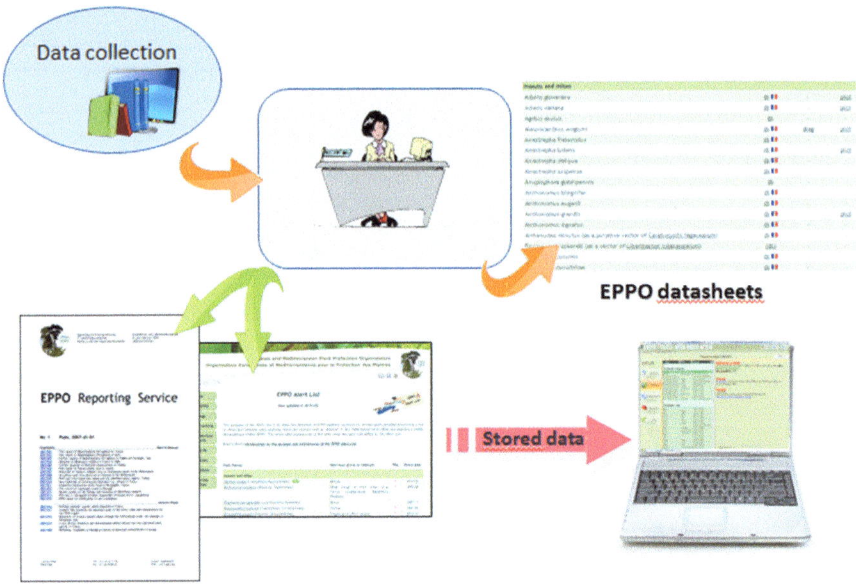

Fig. 3.2 Illustration of the information flow within EPPO

2.5.1 The PQR Database

PQR (Plant Quarantine data Retrieval system) provides detailed information on the geographical distribution and host plants of quarantine pests. Its search tools also allow the users to identify commodities which are able to act as pathways in international trade for the movement of pests and diseases. In recent years, the database has been extended to cover invasive alien plants. The PQR version also includes world maps, pictures and active links to the EPPO Reporting Service (see Fig. 3.3). PQR can be downloaded free of charge from the EPPO website.

2.6 Participation in the International Standard Setting Process

EPPO as an RPPO is playing an active role in the international Standard setting process within the IPPC. This participation involves making suggestions for topics for ISPMs but also coordinating the preparation of regional comments on these draft international Standards.

Regarding seeds, a Standard is in preparation on the international movement of seeds. This Standard was considered necessary to provide additional guidance on phytosanitary measures related to the international movement of seed. Indeed,

Fig. 3.3 PQR screenshots

important pests are seed-transmissible and the movement of infested seed may pose a risk for the international spread of pests.

The Scope and purpose of the Standard as given in the specification (see https://www.ippc.int/publications/specification-54-international-movement-seed) is presented below.

Scope and Purpose

This Standard would apply to seed moved internationally (including forest tree seeds). The proposed Standard is intended to provide additional guidance to assist NPPOs to identify, assess and manage the pest risk associated with the international movement of seed. The Standard may also facilitate the international movement of seed through increased harmonization of phytosanitary import requirements. It should identify and describe specific phytosanitary measures that could be used to reduce pest risk associated with the international movement of seed, including phytosanitary measures that may be applied during growth, at seed harvest, seed extraction, during post-harvest seed processing, and on arrival, inspection and testing. The Standard would not apply to grain. This Standard will help minimize the risk of the global spread of pests of plants including those which can be considered invasive alien species and other organisms whose pest risk has not yet been identified.

Experts from the EPPO region have participated in the drafting and reviewing of this Standard which is currently under member consultation.

3 Conclusions

As international trade continues to increase, EPPO countries have an increasingly difficult task to protect their environment and crops. With horizon scanning activities, risk analysis, communication and harmonization of plant pest diagnostics, EPPO provides a major contribution to the prevention of introduction of new pests from other parts of the world, and to the limitation of their spread within the region should they be introduced.

References

Baker RHA (2012) An introduction to the PRATIQUE Research Project. EPPO Bull 42:1–2. doi:10.1111/j.1365-2338.2011.02522.x

EPPO (2007) PM 7/84 (1) Basic requirements for quality management in plant pest diagnosis laboratories. EPPO Bull 37:580–588

EPPO (2009) PM 9/10(1): generic elements for contingency plans. EPPO Bull 39:471–474. doi:10.1111/j.1365-2338.2009.02332.x

EPPO (2010) PM 7/98 (1) specific requirements for laboratories preparing accreditation for a plant pest diagnostic activity. OEPP Bull 40:5–22

EPPO (2011) Standard on PRA PM 5/3(5). Decision-support scheme for quarantine pests. http://archives.eppo.int/EPPOStandards/pra.htm. Accessed 20 Aug 2013

EPPO (2012) Decision-support scheme for an Express Pest Risk Analysis. EPPO Bull 42:457–462. doi:10.1111/epp.2591

Griessinger D, Suffert M, Brunel S, Petter F (2012) CAPRA: the EPPO computer assisted PRA scheme. EPPO Bull 42:42–47. doi:10.1111/j.1365-2338.2012.02541.x

IPPC (2013) Pest risk analysis for quarantine ISPM no. 11 in international standards for phytosanitary measures. IPPC Secretariat, FAO, Rome, pp 135–160

MacLeod A, Anderson H, Follak S, van der Gaag DJ, Potting R, Pruvost O, Smith J, Steffek R, Vloutoglou I, Holt J, Karadjova O, Kehlenbeck H, Labonne G, Renaud P, Viaene N, Anthoine G, Holeva M, Hostachv B, Ilieva Z, Karssen G, Krumov V, Limon P, Meffert J, Niere B, Petrova E, Peyre J, Pfeilstetter E, Roelofs W, Rothlisberger F, Sauvion N, Schenck N, Schrader G, Schroeder T, Steinmöller S, Tjou-Tam-Sin L, Ventsislavov V, Verhoeven K, Wesemael W (2012) Pest risk assessment for the European Community plant health: a comparative approach with case studies. Supporting Publications 2012:EN-319 [1053 pp]. Available online www.efsa.europa.eu/publications

Petter F, Suffert M (2010) Survey on the use of tests mentioned in EPPO diagnostic protocols. OEPP BULL/EPPO BULL 40:121–126

Petter F, Brunel S, Suffert M (2010) Pest risk analysis as applied to plant pathogens. In: Strange RN, Gullino ML (eds) The role of plant pathology in food safety and food security, vol 3, Plant pathology in 21st century. Springer, Dordrecht. doi:10.1007/978-1-4020-8932-9_12

Petter F, Suffert M, Roy AR, Griessinger D, McMullen M (2013) The European and Mediterranean Plant Protection Organization, one of our objectives: serving the needs of plant pest diagnostic laboratories. EUROREFERENCE 9. http://www.ansespro.fr/euroreference/Documents/ER9-FocusEN.pdf. Accessed 27 Sept 2013

Roumagnac P, Pruvost O, Chiroleu F, Hughes G (2004) Spatial & temporal analyses of bacterial blight of onion caused by Xanthomonas axonopodis pv. allii. Phytopathology 9:138–146

Roy AS, Petter F, Griessinger D (2010) EPPO database on diagnostic expertise: http://dc.eppo.org. EPPO Bull 40:127–130

Zlof V, Smith IM, McNamara DG (2000) Protocols for the diagnosis of quarantine pests. EPPO Bull 30:361–363

Chapter 4
Seed-Borne Fungal Pathogens of Leafy Vegetable Crops

M. Lodovica Gullino, Giovanna Gilardi, and Angelo Garibaldi

Abstract Leafy vegetables are economically important crops, with a relevant role in the diet all over the world, grown worldwide under intensive cultivation systems. In the past few years, in coincidence with the intensification of such cultivations, many new pathogens emerged, causing severe losses. Many of them are seed-borne and their transmission through infected seeds guarantees their rapid spread in different geographic areas. A relatively small percent of contaminated seeds is often sufficient to cause high disease incidence. The leafy vegetable sector is particularly exposed to the risk of the emergence of new diseases as a consequence of its dynamism, the wide range of products, continual innovation in procedures or in products and the use of intensive cultivation techniques that characterize it. Italy, with its very intensive vegetable production, represents indeed a very interesting case study. This chapter will review the situation observed in different production areas, with special reference to Italy. The crops considered are lettuce, wild and cultivated rocket, lamb's lettuce, chicory, endive, basil and spinach with particular regard to the Fusarium and Verticillium wilt agents as well as emerging leaf pathogens (*Alternaria* spp., *Plectosphaerella cucumerina*, downy mildew agents, *Cladosporium variabile* and *Stemphylium botryosum*).

Keywords Vegetable production • Pathogen's spread • Lettuce • Disease management

1 Introduction

Leafy vegetables are economically important crops, with a relevant role in the diet all over the world. The development and success of ready-to-eat processed preparations based on single or mixed leafy vegetables increased the interest for

M.L. Gullino (✉)
AGROINNOVA, University of Torino, Largo Braccini 2, 10095 Grugliasco, Italy

DISAFA, University of Torino, Largo Braccini 2, 10095 Grugliasco, Italy
e-mail: marialodovica.gullino@unito.it

G. Gilardi • A. Garibaldi
AGROINNOVA, University of Torino, Largo Braccini 2, 10095 Grugliasco, Italy

M.L. Gullino, G. Munkvold (eds.), *Global Perspectives on the Health of Seeds and Plant Propagation Material*, Plant Pathology in the 21st Century 6,
DOI 10.1007/978-94-017-9389-6_4

Table 4.1 Contamination of lettuce, endive, chicory, rocket, corn salad, spinach, and basil seeds by some fungal pathogens (Modified from Gullino et al. 2012)

Crop	Pathogen	% of infected seeds	Reference
Lettuce	*Fusarium oxysporum* f. sp. *lactucae*	0.1	Garibaldi et al. 2004a
Lettuce	*Verticillium dahliae*	66–90	Vallad et al. 2005
Lettuce	*Botrytis cinerea*	30	Sowley et al. 2010
Lettuce	*Microdochium panattonianum*	[a]	Sutton and Holderness 1986
Endive, chicory, escarole	*Alternaria cichorii*	0.6–13.75	Barreto et al. 2008
Endive and chicory	*Microdochium panattonianum*	[a]	Sutton and Holderness 1986
Rocket	*Fusarium oxysporum*	0.1	Garibaldi et al. 2004b
Wild rocket	*Plectosphaerella cucumerina*	0.15	Gilardi et al. 2013a
Corn salad	*Phoma valerianellae*	0.6–15	Pellegrino et al. 2010
Spinach	*Fusarium oxysporum* f. sp. *spinaceae*	[a]	Bassi and Goode 1978
Spinach	*Peronospora farinosa* f. sp. *spinaciae* races 1,2,3,4	0.3–2.9	Lorenzini and Nali 1994; Inaba et al. 1983
Spinach	*Cladosporium variabile*	1.8	Matta and Garibaldi 1981; Hernandez-Perez and du Toit 2006
Spinach	*Stemphylium botryosum*	1–95	Hernandez-Perez and du Toit 2006
Spinach	*Verticillium dahliae*	0.3–84.8	du Toit et al. 2005
Basil	*Fusarium oxysporum* f. sp. *basilici*	0.4 (external contamination)	Martini and Gullino 1991
		0.2 (embrio contamination)	
Basil	*Alternaria alternata*	0.2–15.0 (external contamination)	Gilardi et al. 2013b
		0.2–2.0 (embrio contamination)	
Basil	*Peronospora belbahrii*	0.01	Garibaldi et al. 2004d

[a]Data not available

such crops. In different geographic areas worldwide intensive cultivation systems have been successfully exploited for leafy vegetable crops. In coincidence with the intensification of such cultivations, many new pathogens emerged, causing severe losses. Many of them resulted seed-borne and their transmission through infected seeds guarantees their rapid spread in different geographic areas. Often, a relatively small percent of contaminated seeds leads to high disease incidence (Table 4.1). The leafy vegetable sector is particularly exposed to the risk of the emergence of new diseases as a consequence of its dynamism, the wide range of products, continual innovation in procedures or in products and the use of intensive cultivation techniques that characterize it. Italy, with its very intensive vegetable production, represents indeed a very interesting case study. The sudden, and almost contemporary, appearance of new diseases on leafy vegetables grown in different continents can be traced back to structural reasons. In this line of production, in fact, the propagation material is produced in just a few big nurseries, which in turn supply small nurseries in other regions or countries (Garibaldi and Gullino 2010). To reduce the risk of spreading new diseases, it would be necessary to intercept the pathogens in the nodal points of the line.

This chapter will review the situation observed in different production areas, with special reference to Italy, a country where the cultivation of leafy vegetables for ready-to-eat products is very important and constantly increasing, during the past 10 years. The crops considered are lettuce (*Lactuca sativa* L.), wild (*Diplotaxis* spp.) and cultivated (*Eruca sativa* Mill.) rocket, lamb's lettuce (*Valerianella olitoria* L.), chicory (*Cichorium intybus*), endive (*Cichorium endivia* L.), basil (*Ocimum basilicum* L.) and spinach (*Spinacia oleracea* L.).

2 Fusarium Wilts

Different *formae speciales* of *Fusarium oxysporum* affect leafy vegetables, representing a potential threat to their production in many areas. This pathogen possess exceptional mechanisms for survival and dissemination. Seed transmission occurs when propagules are carried as surface or internal contaminants of seeds or in associated plant debris. Many Fusarium wilts of leafy vegetables, such as those of lettuce, rocket and basil possess these traits.

The Fusarium wilt caused by *Fusarium oxysporum* f. sp. *lactucae* was first identified in 1955 as the cause of a root rot on lettuce in Japan (Matuo and Motohashi 1967). Some 35 years later, a Fusarium wilt was reported in 1990 on lettuce in the United States (California) and the causal pathogen was named *Fusarium oxysporum* f. sp. *lactucum* (Hubbard and Gerik 1993). Later research demonstrated that the California pathogen and Japanese race 1 belonged to the same compatibility group and were considered to be the same *forma specialis* (Fujinaga et al. 2003). Subsequent recognition of the pathogen on lettuce has been reported in Iran in 1995, Taiwan in 1998, Brazil in 2000, Italy in 2002 and in the state of Arizona in the United States in 2001 (Matheron and Gullino 2012).

Fusarium wilts also have been observed on several salad crops in addition to lettuce, as reviewed by Matheron and Gullino (2012). A wilt incited by *F. oxysporum* on *Cichorium endivia* was observed in 2007 in northern Italy (Garibaldi et al. 2009). A wilt of *E. sativa* attributed to *F. oxysporum* f. sp. *erucae* was reported in India in 1973 and 1987. In 2001 and 2002, Fusarium wilt developed in northern Italy on wild and cultivated rocket (Matheron and Gullino 2012). *F. oxysporum* f. sp. *conglutinans* and *F. oxysporum* f. sp. *raphani* are the causal agents of wilt of wild and cultivated rocket (Garibaldi et al. 2006; Catti et al. 2007). In 2003, a new wilt was observed on lamb's lettuce in northern Italy, on the cvs. Trophy and Palmares, incited by *F. oxysporum* f. sp. *conglutinans* (Matheron and Gullino 2012). This wilt developed in the same area where Fusarium wilts of lettuce, wild and cultivated rocket were previously observed. On endive, *F. oxysporum* causes stunting and yellowing of affected plants, which also show a poorly developed root system (Garibaldi et al. 2009). On chicory, Fusarium affected plants were chlorotic and stunted, with poorly developed root system. Black streaks were observed in the stem and proximal part of the leaf vascular system in wilted plants (Garibaldi et al. 2011c). The causal agent has been identified as a new *forma specialis*, *F. oxysporum* f. sp. *cichorii* (Poli et al. 2012).

The appearance of Fusarium wilt on lettuce in geographically distant areas, such as Brazil, Iran, Italy, Taiwan, and the United States, at least 35 years after the initial discovery of this disease in Japan, suggests a long-distance method of dispersal of *F. oxysporum* f. sp. *lactucae*. Seed transmission of the pathogen is a possible dissemination mechanism. Garibaldi et al. (2004a) discovered that 9 of 27 samples of lettuce seed obtained from commercial seed lots planted in fields, that were subsequently affected by Fusarium wilt in Italy, were contaminated by *F. oxysporum*. Also, *F. oxysporum* f. sp. *conglutinans* and *F. oxysporum* f. sp. *raphani*, causal agents of Fusarium wilt of wild and cultivated rocket, are seed-transmitted (Garibaldi et al. 2004b). Therefore, seed transmission on wild and cultivated rocket seeds contributed to the spread of the disease in Italy. Other means of pathogen dispersal within and between fields would include any farming operation that would move infested soil or plant material, such as seed-bed preparation activities, cultivation, movement of mud-encrusted sprinkler-irrigation pipe, and harvesting crew operations.

On basil, *F. oxysporum* f. sp. *basilici*. was first described on basil in the former USSR and later spread to many basil growing areas (Garibaldi et al. 1997), causing severe damages also due to its soil-and airborne behaviour (Gamliel et al. 1996). The pathogen has been isolated from seeds, before and after disinfestation with sodium hypochlorite: Martini and Gullino (1991) found that 0.4 % of non-disinfested and 0.2 % of disinfested commercial seeds harbored *F. oxysporum* f. sp. *basilici*. It is not known whether *F. oxysporum* f. sp. *basilici* is an external contaminant or infects seeds internally (Martini and Gullino 1991; Vannacci et al. 1999), although diseased plants have been obtained from some seed lots after external disinfection (Vannacci et al. 1999). Epidemiological considerations suggest that rapid local spread of Fusarium wilt and crown rot of basil is caused by airborne inoculum derived mainly from macroconidial masses on stem surfaces (Gamliel et al. 1996), through soil particles, and during harvest, whereas seed-borne inoculum is probably responsible for its long-distance transmission (Martini and Gullino 1991; Elmer et al. 1994; Gamliel et al. 1996; Elmer 2001).

Also *F. oxysporum* f. sp. *spinaciae*, causal agent of Fusarium wilt on spinach is seed-borne (Bassi and Goode 1978).

3 Verticillium Wilt

Caused by *Verticillium dahliae*, this disease has been observed on lettuce, chicory and spinach (Ciccarese et al. 1987; Correll et al. 1994; Davis et al. 1997; Garibaldi et al. 2007). It is important in the presence of air and soil temperatures of 20–25 °C, causing more losses during spring and fall. In the case of lettuce, seed transmission of the pathogen plays an important role: Vallad et al. (2005) reported a very high percentage (66–90 %) of infected seeds. In the case of spinach, seed contamination has been proved. *V. dahliae* is systemic in spinach and readily seed transmitted (Du Toit et al. 2005). On spinach, the spread of this pathogen throughout infected seeds is at present a major concern in areas where fresh and processed spinach crops are grown in rotation with other crops susceptible to the pathogen (Maruthachalam et al. 2013).

4 Foliar Diseases

Downy mildew of basil, incited by *Peronospora belbahrii*, (Belbahri et al. 2005; Thines et al. 2009) was observed in northern Italy in 2003 (Garibaldi et al. 2004c) and quickly spread to other Italian regions in Central and Southern Italy (Garibaldi and Gullino 2010) as well as France (Garibaldi et al. 2005). This pathogen was first reported in Uganda, identified as *Peronospora* sp. (Hansford 1933) and much later in Switzerland (Lefort et al. 2003). After this report in Switzerland, the pathogen spread to many basil growing areas. The disease was recently observed also in Belgium (Coosemans 2004), in the USA (Roberts et al. 2009), in Cuba (Martinez de La Parte et al. 2010) and in Hungary (Nagy and Horváth 2011). Its spread probably has been favored by the fact that it is seed-transmitted (Garibaldi et al. 2004d).

Also *Peronospora farinosa* f. sp. *spinaciae* (syn. *P. effusa*), causing downy mildew of spinach¸ is transmitted throughout seeds. Already in 1935, Cook reported that spinach crops grown from heavily infested seeds bearing oospores were severely damaged by downy mildew. Inaba et al. (1983) showed that the percentage of spinach seedlings infected by downy mildew was positively correlated with the degree of oospore infestation of seeds.

Phoma valerianellae, the causal agent of a foliar disease of lamb's lettuce, is another seed-borne pathogen (Nathaniels 1985). Its recent resurgence and spread in Italy in areas devoted to ready-to-eat production has been explained with such characteristics (Pellegrino et al. 2010).

The recent outbreak of *Plectosphaerella cucumerina* on wild rocket represents a potential threat to rocket production in Italy as well as elsewhere. The disease has been detected on wild rocket, widely grown for processing (Garibaldi et al. 2012). *P. cucumerina*, is frequently seed-transmitted (four seed samples out of eight tested were contaminated), which suggests that seeds may be important in disseminating this pathogen, despite a low level of contamination (about 0.15 %) in the tested samples (Gilardi et al. 2013a). The fast spreading of the disease that occurred first in southern Italy in 2012, moving in a few months to northern Italy (Gilardi et al. 2012) can be explained with the capability of the pathogen to infect seeds. The pathogen was recently detected also on endive (Garibaldi et al 2013).

Leaf spot of escarole, chicory and endive, caused by *Alternaria cichorii*, is easily transmitted by infected seeds (Barreto et al. 2008).

A leaf spot of basil, causing extensive necrosis and incited by *Alternaria* spp., appeared recently in several countries. Taba et al. (2009) showed that the black lesion of basil grown in greenhouse in Japan were caused by *Alternaria alternata*. Recently in Israel, a similar black spot caused by *Alternaria* sp. was observed at the harvesting of summer basil (Kenigsbuch et al. 2010). A similar leaf spot was observed during the summer-fall 2010 on sweet basil, grown in soilless systems as well as in soil in northern Italy (Garibaldi et al. 2011a). All 18 Italian seed samples tested resulted contaminated by *Alternaria* spp. The frequency of isolation of *Alternaria* spp. colonies was higher in the case of not disinfected seeds for all samples tested. For instance, in the case of seeds belonging to experimental lines of basil, the frequency of isolation of *Alternaria* spp. from seeds was 1.18 % for not disinfected seeds and 0.43% for disinfected seeds. In the case of seeds belonging to commercial varieties of basil, *Alternaria* spp. was isolated respectively from 7.29 % to 2.62 % of not disinfected and disinfected seeds (Gilardi et al. 2013b). *Alternaria japonica* was recently reported as the cause of a new leaf spot on wild and cultivated rocket (Garibaldi et al. 2011b).

Cladosporium variabile and *Stemphylium botryosum*, causal agents of two leaf spots of spinach, are booth seed-borne In the case of *S. botryosum*, the presence of the pathogens in seed lots, combined with international movement of spinach seeds, might explain the sudden and almost concomitant appearance of the pathogen in several states of the USA (Hernandez-Perez and du Toit 2006).

Also *Botrytis cinerea*, causal agent of grey mould of lettuce, is often present in symptomless lettuce plants as a systemic, endophytic, infection which may arise from seeds (Sowley et al. 2010).

Concluding Remarks

A high number of fungal pathogens causing severe losses in leafy vegetables is seed-borne. These characteristics strongly influences their easy and rapid spread in several production areas from the point of origin. The fact that seed production is very much concentrated in few establishments, in the case of their contamination during the production process, favors the quick spread of new diseases throughout seed commercialization.

This phenomenon happened many times in the past 15–20 years, causing the rapid spread of new pathogens in many production areas. Italy, as shown

(continued)

by the many new reported above described, represents an interesting case study. Indeed, from one side, leafy vegetable production is quite important in terms of cultivated surfaces, with a continuous intensification of the cultural systems, more and more devoted to ready-to-eat production. From the other side, most seeds used are imported and the use of seeds contaminated has been proved many times. This fact also bring under our attention the need for training more researchers and experts in the field of seed pathology as well as for strengthening the relationships between academia and industry (Munkvold 2009).

Identifying the primary source of inoculum is of critical importance for effective disease management, as reviewed by Lievens et al. (2012) and Gullino et al. (2014b) as well as developing effective control methods (Gullino et al. 2014a; Kock and Roberts 2014).

Acknowledgements The results reported have received funding from the European Union's Seventh Framework Programme (FP7/2007–2013) under grants agreement n°261752, "PlantFoodSec – Plant and Food Biosecurity, Network of Excellence" and n°311875 "TESTA – Seed health: development of seed treatment methods, evidence for seed transmission and assessment of seed health".

References

Barreto RW, Santin AM, Vieira BS (2008) *Alternaria cichorii* in Brazil on *Cichorium* spp. seeds and cultivated and weedy hosts. J Phytopathology 156:425–430

Bassi A, Goode MJ (1978) *Fusarium oxysporum* f. sp. *spinaceae* seedborne in spinach. Plant Dis Repor 62:203–205

Belbahri I, Calmin G, Pawlowski J, Lefort F (2005) Phylogenetic analysis and real time PCR detection of a presumably undescribed *Peronospora* species on sweet basil and sage. Mycol Res 109:1276–1287

Catti A, Pasquali M, Ghiringhelli D, Garibaldi A, Gullino ML (2007) Analysis of vegetative compatibility groups of *Fusarium oxysporum* from *Eruca vesicaria* and *Diplotaxis tenuifolia*. J Phytopathol 155:61–64

Ciccarese F, Frisullo S, Cirulli M (1987) Severe outbreaks of Verticillium wilt on *Cichorium intybus* and *Brassica rapa* and pathogenic variations among isolates of *Verticillium dahliae*. Plant Dis 71:1144–1145

Cook HT (1935) Occurrence of oospores of *Peronospora effusa* with commercial spinach seed. Phytopathology 25:11–12

Coosemans J (2004) First report of *Peronospora lamii*, downy mildew on basil (*Ocimum basilicum*) in Belgium. Parasitica 60:27

Correll JC, Morelock TE, Black MC, Koike ST, Brandenberger LP, Dainello FJ (1994) Economically important diseases of spinach. Plant Dis 78:653–660

Davis RM, Subbarao KV, Raid RN, Kurtz EA (1997) Compendium of lettuce diseases. American Phytopathological Society Press, St. Paul, p 79

Du Toit LJ, Derie ML, Hernandez-Perez P (2005) Verticillium wilt in spinach seed production. Plant Dis 89:4–11

Elmer WH (2001) Seeds as vehicles for pathogen importation. Biol Invasions 3:263–271
Elmer WH, Wick RL, Haviland P (1994) Vegetative compatibility among *Fusarium oxysporum* f. sp. *basilicum* isolates recovered from basil seed and infected plants. Plant Dis 78:789–791
Fujinaga M, Ogiso H, Tuchiya N, Saito H, Yamanaka S, Nozue M, Kojima M (2003) Race 3, a new race of *Fusarium oxysporum* f. sp. *lactucae* determined by differential system with commercial cultivars. J Gen Plant Pathol 69:23–28
Gamliel A, Katan T, Yunis H, Katan J (1996) Fusarium wilt and crown rot of sweet basil: involvement of soilborne and airborne inoculum. Phytopathology 86:56–62
Garibaldi A, Gullino ML (2010) Emerging soilborne diseases of horticultural crops and new trends in their management. Acta Hortic 883:37–46
Garibaldi A, Gullino ML, Minuto G (1997) Diseases of basil and their management. Plant Dis 81:124–132
Garibaldi A, Gilardi G, Gullino ML (2004a) Seed transmission of *Fusarium oxysporum* f. sp. *lactucae*. Phytoparasitica 32:61–65
Garibaldi A, Gilardi G, Pasquali M, Keiji S, Gullino ML (2004b) Seed transmission of *Fusarium oxysporum* of *Eruca vesicaria* and *Diplotaxis muralis*. J Plant Dis Prot 111:345–350
Garibaldi A, Minuto A, Minuto G, Gullino ML (2004c) First report of downy mildew of basil (*Ocimum basilicum*) in Italy. Plant Dis 88:312
Garibaldi A, Minuto G, Bertetti D, Gullino ML (2004d) Seed transmission of *Peronospora* sp. of basil. J Plant Dis Prot 111:465–469
Garibaldi A, Minuto A, Gullino ML (2005) First report of downy mildew caused by *Peronospora* sp. on basil (*Ocimum basilicum*) in France. Plant Dis 89:683
Garibaldi A, Gilardi G, Gullino ML (2006) Evidence for an expanded host range of *Fusarium oxysporum* f. sp. *raphani*. Phytoparasitica 34:115–121
Garibaldi A, Gilardi G, Gullino ML (2007) First report of Verticillium wilt caused by *Verticillium dahliae* on lettuce in Italy. Plant Dis 91:770
Garibaldi A, Gilardi G, Troisi M, Gullino ML (2009) First report of Fusarium wilt of endive (*Cichorium endivia*) caused by *Fusarium oxysporum* in Italy. Plant Dis 93:1078
Garibaldi A, Gilardi G, Bertoldo C, Gullino ML (2011a) First report of leaf spot of sweet basil (*Ocimum basilicum*) caused by *Alternaria alternata* in Italy. J Plant Pathol 93(S4):71
Garibaldi A, Gilardi G, Bertoldo C, Gullino ML (2011b) First report of leaf spot of wild (*Diplotaxis tenuifolia*) and cultivated (*Eruca vesicaria*) rocket caused by *Alternaria japonica* in Italy. Plant Dis 95:1316
Garibaldi A, Gilardi G, Poli A, Gullino ML (2011c) First report of Fusarium wilt of chicory (*Cichorium intybus*) caused by *Fusarium oxysporum* in Italy. Plant Dis 95:496
Garibaldi A, Gilardi G, Ortu G, Gullino ML (2012) First report of *Plectosphaerella cucumerina* on greenhouse cultured wild rocket (*Diplotaxis tenuifolia*) in Italy. Plant Dis 96:1825
Garibaldi A, Gilardi G, Ortu G, Gullino ML (2013) First report of a new leaf spot caused by *Plectosphaerella cucumerina* on field grown endive (*Cichorium endivia*) in Italy. Plant Dis 97:848
Gilardi G, Bertoldo C, Gullino ML, Garibaldi A (2012) Una nuova malattia della rucola coltivata e selvatica causata da *Alternaria japonica* in Italia. Protezione delle Colture 5(1):26–28
Gilardi G, Garibaldi A, Gullino ML (2013a) Seed transmission of *Plectosphaerella cucumerina*, causal agent of leaf spot of *Diplotaxis tenuifolia* in Italy. Phytoparasitica 41:411–416
Gilardi G, Gullino ML, Garibaldi A (2013b) Occurrence of *Alternaria* spp. in seeds on basil and its pathogenicity. J Plant Pathol 95:41–47
Gullino ML, Gilardi G, Garibaldi A (2012) Aumentano i rischi di diffusione dei patogeni mediante seme infetto: l'esempio delle piante orticole da foglia. Protezione delle Colture 5(2):14–18
Gullino ML, Gilardi G, Garibaldi A (2014a) Chemical and non-chemical seed dressing for leafy vegetable crops. In: Gullino ML, Munkvold G (eds) Global perspectives on the health of seeds and plant propagation material. Springer, Dordrecht, pp 125–136
Gullino ML, Gilardi G, Ortu G, Garibaldi A (2014b) Development and implementation of rapid and specific detection techniques for seed-borne pathogens of leafy vegetable crops. In:

Gullino ML, Bonants P (eds) Detection and diagnostics of plant pathogens. Springer, Dordrecht (in press)

Hansford CG (1933) Annual report of the mycologist. Rev Appl Mycol 12:421–422

Hernandez-Perez P, du Toit LJ (2006) Seedborne *Cladosporium variabile* and *Stemphylium botryosum* in spinach. Plant Dis 90:137–145

Hubbard JC, Gerik JS (1993) A new wilt disease of lettuce incited by *Fusarium oxysporum* f. sp. *lactucum* forma specialis nov. Plant Dis 77:750–751

Inaba T, Takahashi K, Morinaka T (1983) Seed transmission of spinach downy mildew. Plant Dis 74:1139–1141

Kenigsbuch D, Chalupowicz D, Aharon Z, Maurer D, Ovadia A, Aharoni N (2010) Preharvest solar heat treatment for summer basil (*Ocimum basilicum*) affects decay during shipment and shelf life. Acta Hortic 880:161–166

Kock E, Roberts SE (2014) Non-chemical seed treatment in the control of seed-borne pathogens. In: Gullino ML and Munkvold G (eds) Global perspectives on the health of seeds and plant propagation material. Springer, Dordrecht, pp 105–123

Lefort F, Gigon V, Amos B (2003) Le mildiou s'étend. Dé jà détectédans des nombreux pays européens, *Peronospora lamii*, responsible du mildiou de basilic, a éte´ observe en Suisse dansla region lémanique. Réussir Fruits et Légumes 223:66

Lievens B, Hanssen IM, Rep M (2012) Recent developments in the detection and identification of *formae speciales* and races of *Fusarium oxysporum*: from pathogenicity testing to molecular diagnosis. In: Gullino ML, Katan J, Garibaldi A (eds) Fusarium wilts of greenhouse crops. APS Press, ST Paul, pp 47–55

Lorenzini G, Nali C (1994) A new race (race 4) of spinach downy mildew in Italy. Plant Dis 78:208

Martinez de La Parte E, Pérez-Vicente L, Bernal B, Garcí D (2010) First report of *Peronospora* sp. on sweet basil (*Ocimum basilicum*) in Cuba. Plant Pathol 59:800

Martini P, Gullino ML (1991) Trasmissibilità per seme di *Fusarium oxysporum* f. sp. *basilicum* agente della tracheofusariosi del basilico. Informatore Fitopatologico 41(9):59–61

Maruthachalam K, Klosterman SJ, Ancheta A, Mou BQ, Subbarao KV (2013) Colonization of spinach by *Verticillium dahliae* and effects of pathogen localization on the efficacy of seed treatments. Phytopathology 103:268–280

Matheron M, Gullino ML (2012) Fusarium wilt of lettuce and other salad crops. In: Gullino ML, Katan J, Garibaldi A (eds) Fusarium wilt of greenhouse vegetable and ornamental crops. APS Press, St. Paul, pp 175–183

Matta A, Garibaldi A (eds) (1981) Malattie delle piante ortensi. Edagricole, Bologna

Matuo T, Motohashi S (1967) On *Fusarium oxysporum* f. sp. *lactucae* n. f. causing root rot on lettuce. Trans Mycol Soc Jpn 32:13–15

Munkvold GP (2009) Seed pathology progress in academia and industry. Annu Rev Plant Physiol Plant Mol Biol 47:285–311

Nagy G, Horváth A (2011) Occurrence of downy mildew caused by *Peronospora belbahrii* on sweet basil in Hungary. Plant Dis 95:1034

Nathaniels NQR (1985) *Phoma valerianellae* on corn salad. Plant Pathol 34:449–450

Pellegrino C, Gilardi G, Gullino ML, Garibaldi A (2010) Detection of *Phoma valerianellae* in lamb's lettuce seeds. Phytoparasitica 38:159–165

Poli A, Gilardi G, Spadaro D, Gullino ML, Garibaldi A (2012) Molecular characterization of *Fusarium oxysporum* f. sp. *cichorii* pathogenic on chicory (*Cichorium intybus*). Phytoparasitica 40:383–391

Roberts PD, Raid RN, Harmon PF, Jordan SA, Palmateer AJ (2009) First report of downy mildew caused by a *Peronospora* sp. on basil in Florida and the United States. Plant Dis 93:199

Sowley ENK, Dewey FM, Shaw MW (2010) Persistent, symptomless, systemic and seed-borne infection of lettuce by *Botrytis cinerea*. Eur J Plant Pathol 126:61–71

Sutton BC, Holderness M (1986) *Microdochium panattonianum* IMI Descriptions of Fungi and bacteria. Set 104. Trans Br Mycol Soc 86:620

Taba S, Takara A, Nasu K (2009) Alternaria leaf spot of basil caused by *Alternaria alternata* in Japan. J Gen Plant Pathol 75:160–162

Thines M, Telle S, Ploch S, Runge F (2009) Identity of the downy mildew pathogens of basil, coleus and sage with implications for quarantine measures. Mycol Res 113:532–540

Vallad GE, Bhat RG, Koike ST, Ryder E, Subbarao KV (2005) Weed reservoirs and seed transmission of *Verticillium dahliae* in lettuce. Plant Dis 89:317–323

Vannacci G, Cristiani C, Forti M, Kontoudakis G, Gambogi P (1999) Seed transmission of *Fusarium oxysporum* f. sp. *basilici* in sweet basil. J Plant Pathol 81:47–53

Part II
Detection

Chapter 5
Technical Challenges for Specific, Sensitive Detection of Seed-Borne Bacterial Pathogens

Norman W. Schaad, Carmen N. Mortensen, Jianqiang Li, Jianjun Feng, Laixin Luo, Angelo Mazzaglia, and Giorgio M. Balestra

Abstract Seed-borne pathogens are a major threat to agriculture production and security in a fast-moving global economy. As global trade increases so does the threat of accidental and deliberate introduction of seed-borne pathogens. Seed transmitted diseases can result in severe economic losses. The challenge is for government and industry to cooperate in providing pathogen-free seeds. Seeds can be assayed and infested lots destroyed or treated by chemical and/or physical means and re-assayed. Considerable progress has been made in developing reliable sensitive and specific assays. However, technical challenges remain. Seeds are challenging because they are often heavily contaminated with saprophytic bacteria. This makes agar plating difficult and often inhibits molecular-based protocols such as PCR. Use of DNA sequence information for designing highly specific and sensitive PCR primers is especially challenging. To avoid false negative results in classical PCR, internal controls of primers targeting bacterial 16S rDNA or plant 26S mitochondrial rDNA can be applied. Perhaps the most reliable and sensitive PCR protocol is real-time PCR using probe-based protocols such as TaqMan. The biggest challenge to PCR has been problems with PCR inhibitors present in seeds. These inhibitors can be partially removed using such treatments as heat,

N.W. Schaad (✉)
USDA, ARS, Foreign Disease Weed Science-RU, Ft. Detrick, MD, USA

2414 Rikkard, Thousand Oaks, CA, USA
e-mail: lionschaad@gmail.com

C.N. Mortensen
Danish Seed Health Centre, Department of Plant and Environmental Sciences, University of Copenhagen, Copenhagen, Denmark

3450 Havrevaenget 4, Alleroed, Denmark

J. Li • L. Luo
Department Plant Pathology, China Agricultural University, Beijing, People's Republic of China

J. Feng
Animal & Plant Quarantine Technology Center, Shenzhen, People's Republic of China

A. Mazzaglia • G.M. Balestra
Department Agriculture and Forestry, Nature and Energy, University of Tuscia, Tuscia, Italy

M.L. Gullino, G. Munkvold (eds.), *Global Perspectives on the Health of Seeds and Plant Propagation Material*, Plant Pathology in the 21st Century 6,
DOI 10.1007/978-94-017-9389-6_5

DNA extraction, immunocapture, and BIO-PCR. BIO-PCR not only reduces inhibitors but greatly increases sensitivity by allowing the target bacterium to multiply. This expands our ability to detect low numbers of pathogens even in seeds contaminated with large numbers of saprophytes. A major challenge for plant quarantine is dealing with PCR positive results without cultures for confirmation.

Keywords Seed-borne • Bacteria • Detection • Polymerase chain reaction

1 Detection of Seed-Borne Pathogens: Regional and Global Solutions

Seed-borne and propagative-borne pathogens are a major threat to agriculture production and security both locally and globally (Schaad et al. 2003). As global trade increases so does the threat of accidental or deliberate introduction of seed-borne pathogens. Seed transmitted diseases can result in severe economic losses to growers. However, infected seeds also provide additional possibilities for reliable controls. By eliminating infested seed lots through quality control, seed health programs, seed transmitted diseases can be prevented (Saettler et al. 1989). The challenge is for government and industry to cooperate in providing pathogen-free seeds. Seeds can be assayed and infested seed lots treated by chemical and/or physical means and re-assayed. Considerable progress has been made over the past 20 or more years in developing protocols for the detection of seed-borne pathogens. The first edition of the APS Manual on Detection of Plant Pathogenic Bacteria in Seed and Planting Material included only 25 pathogens and assays were primarily based on agar plating techniques (Saettler et al. 1989). In contrast the current edition (Fatmi et al. in press) includes assays for 28 bacteria in seed and 20 bacteria in vegetatively propagated crops and contains numerous

Table 5.1 Vegetable disease assays (Fatmi et al. in press)

Pathogen	Host
Clavibacter michiganensis subsp. *michiganenis*	Tomato
Pseudomonas syringae pv. *tomato*	Tomato
Xanthomonas spp.	Tomato
X. hortorum pv. *carotae*	Carrot
Pantoea spp.	Onion
Burkholderia cepacia	Onion
X. campestris pv. *campestris*	Crucifers
Pseudomonas sp.	Crucifers
X. campestris pv. *vitians*	Lettuce
Acidovorax citrulli	Cucurbits
P. syringae pv. *lachrymans*	Cucumber
Pantoea ananatis	Melon

Table 5.2 Cereal disease assays (Fatmi et al. in press)

Pathogen	Host
Pseudomonas syringae pv. *syringae*	Wheat
Xanthomonas translucens	Wheat, Barley, Rye
Pantoea stewartii	Corn
Clavibacter michiganenis subsp. *nebraskensis*	Maize
X. oryzae pv. *oryzae*	Rice
X. oryzae pv. *oryzicola*	Rice
Acidovorax oryzae	Rice
X. campestris pv. *phaseoli*	Beans
P. syringae pv. *syringae*	Beans
X. fuscans subsp. *fuscans*	Beans
P. syringae pv. *phaseolicola*	Beans
Curtobacterium flaccumfaciens subsp. *flaccumfaciens*	Beans
P. syringae pv. *glycinea*	Soybeans
Rhodococcus fasciens	Chickpea
Rathayibacter spp.	Cereals, grasses
C. michiganenis subsp. *insidious*	Alfalfa

immunodiagnostic and molecular protocols (Table 5.1, vegetable seeds; Table 5.2, Cereal, legume, and grass seeds; and Table 5.3, Other planting materials). Several other international organizations such as the International Seed Testing Association (ISTA), International Seed Health Initiatives (ISHI) of ISF (International Seed Federation), European Plant Protection Organization (EPPO), and CAB International have published several assay procedures some of which have been standardized. Assays available include, agar plating, immunodiagnosis, and polymerase chain reaction (PCR). Technical challenges with these assays are numerous and include: avoiding inhibition of growth of target bacteria on semi-selective agar media due to antibiotics produced by saprophytic bacteria; extraction of target bacteria from infected seeds; seed with high numbers of saprophytes, seeds such as those washed in ditch water; and seeds treated with chemicals to eradicate the pathogen. Techniques to extract bacteria from seeds for assaying include: soaking in saline/buffer, soaking under vacuum, wet grinding, soaking and grinding, and enrichment in the seed extract (He and Munkvold 2012). Bacteria in seed and vegetatively- propagated crops provide special challenges because seeds are often harvested and initially cleaned under field conditions and propagated crops are difficult to sample large numbers. Watermelon and melon seeds are commonly washed in the field with ditch water containing high numbers of saprophytic bacteria. Presence of large numbers of saprophytic bacteria makes isolation and identification of the target bacterium difficult on agar media (Alvarez 2004). One novel approach to avoid extraction problems is to combine PCR (see below) with a seedling grow-out assay (Randhawa, unpublished). Many semi-selective agar media have been designed over the past 20 years especially for assaying seeds. Molecular diagnosis is becoming much more prevalent, however, agar plating and immunodiagnostic methods including ELISA, immunofluorescence (IF), lateral

Table 5.3 Planting material assays (Fatmi et al. in press)

Pathogen	Host
Clavibacter michiganenis subsp. *sepedonicus*	Potato
Ralstonia solanacearum	Potato, geranium
Streptomyces scabies	Potato
Liberibacter solanacearum	Potato
Pectobacterium spp.	Potato
Xanthomonas fragariae	Strawberry
L. asiaticus	Citrus
X. citri subsp. *citri*	Citrus
Xylella fastidiosa subsp. *fastidiosa*	Grape, Almond
X. fastidiosa subsp. *pauca*	Citrus
X. fastidiosa subsp. *multiplex*	Peach, shade trees
X. fastidiosa subsp. *sandyi*	Oleander
Erwinia amylovora	Pear, apple, quince
Pseudomonas syringae pv. *actinidae*	Kiwi
P. syringae pv. *savastanoi*	Olive
Agrobacterium spp.	Grape
Leifsonia xyli	Sugarcane
X. albilineans	Sugarcane
Rhodococcus fasciens	Ornamentals
Burkholderia gladioli	Orchids

flow devices, flow cytometry, and immunomagnetic separation (De León et al. 2008) are still used more commonly. Over 90 commercial immunodiagnostic test kits were available as of 2004 (Alvarez 2004). Sensitivity of ELISA is not high being in the range of 10^5–10^6 cfu/ml. Sensitivity can be increased at least 100 times by using IF colony staining (Veena and van Vuurde 2002). Although molecular-based assays are much quicker than agar plating or immunodiagnostic assays and offer higher specificity, many challenges remain. High numbers of saprophytic bacteria can inhibit molecular-based assays such as PCR. Most helpful over the past several years have been technical advances in molecular biology, especially PCR. Between 1989 and 2007, at least 246 papers have been published on use of PCR for detection and identification of plant pathogenic bacteria (Palacio-Bielsa et al. 2009). Molecular information obtained from specific genes and whole genome sequencing offers new challenges for developing highly specific and sensitive assays. The use of sequencing information for designing PCR primers is especially challenging. One major problem is normally only one or two strains have been sequenced. This makes comparative results difficult. Unclear taxonomy is a major problem when determining specificity of an assay. Several molecular-based protocols including classical PCR (Koenraadt et al. 2009; Popović et al. 2010; Umesha et al. 2010), PCR-based denaturing gradient gel electrophoresis (PCR-DGGE) (Hardoim et al. 2012), immunocapture PCR (Guven and Mutlu 2000; Hartung et al. 1996; Walcott and Gitaitis 2000), nested PCR (Bertolini et al. 2003; Hartung et al. 1996; Prosen et al. 1993), real-time PCR (Cho et al. 2012; Schaad et al. 1999; Weller et al. 2000), multiplex PCR (Bertolini et al. 2003; Özdemir 2009; Weller et al. 2000), multiplex nested PCR (Robène-Soustrade et al. 2010), multiplex real-time PCR (Berg et al. 2006; Johnson and Walcott 2012), BIO-PCR (Sakthivel et al. 2001; Schaad et al. 1995, 1999; Song et al. 2004), real-time BIO-PCR (Deuner et al. 2012; Kim et al. 2012; Schaad et al. 1999), membrane real-time BIO-PCR (Schaad et al. 2007), and co-operational PCR (Bertolini et al. 2003) are available for detecting seed-borne bacteria. Co-operational PCR uses three or more primers in a single tube to avoid contamination and improve sensitivity (Bertolini et al. 2003). Multiplex PCR is useful for detecting two or more pathogens in a single assay; however, the challenge is to maintain sensitivity. Chip PCR (microarrays) (Maskos and Southern 1992) utilizing a microchip with DNA probes forming half of the DNA double helix to recognize DNA from a sample being tested is available but has not been developed for seed assays. To avoid false negative results in classical PCR, internal controls of primers targeting bacterial 16S rDNA or plant 26S mitochondrial rDNA should be applied for testing pure bacterial cultures or plant tissue, respectively. Perhaps the most reliable and sensitive PCR-based protocol is real-time PCR (Bach et al. 2003; Schaad et al. 1999, 2003; Weller et al. 2000) using probe-based assays such as Molecular Beacons (Fanelli et al. 2007), Scorpions (De Bellis et al. 2007), or TaqMan (Holland et al. 1991). TaqMan assays are normally favored because they are quite easier to design and are highly sensitive and stable. Several real-time PCR platforms are available including ABI 7700 Sequence Detection System®, (Applied BioSystems, Foster City, CA.); R.A.P.I.D., (Idaho Technology Salt Lake City, Utah); Light Cycler™, Roach Diagnostics Corp. (Indianapolis, IN; SmartCycler®, with 16 chambers (Cepheid, Sunnyvale, CA.); iCycler iq™ from BioRad, Hercules, CA; and Rotor Gene from Corbett, (Sydney, Australia). Results of real-time TaqMan PCR are quantitative and based on fluorescence and a cycle threshold value (Ct) defined as the cycle number at which fluorescence rises above the baseline (Schaad et al. 2003). Challenges to molecular-based assays include, positive results without a viable culture; lack of reliable taxonomic information for choosing positive and negative control strains; designing specific, sensitive PCR primers and probes; a sample size of only 1–5 μl versus 100 μl for agar plating assays; avoiding PCR inhibitors in plant and seeds; maintaining sensitivity in multiplex PCR assays; and redesigning semi-selective agar and liquid media especially for BIO-PCR. Inhibition of real-time PCR is revealed by a delayed or flat Ct curve. These inhibitors can be partially removed using such treatments as polyvinyl-pyrrolidine, heat, DNA extractions, and immunocapture PCR (Palacio-Bielsa et al. 2009; Walcott et al. 2000). Inhibitors can also be avoided by enriching the target bacterium on agar or liquid media before PCR (referred to as BIO-PCR) (Schaad et al. 1995). BIO-PCR is performed as follows: plate 100 μl onto each of six agar plates. A semi-selective medium is preferred. Incubate three plates until pin point-size (micro) colonies can be seen. This is normally 24–48 h, depending on the growth of the target bacterium. The incubation time must be accurately determined as too much time can result in growth of saprophytes and result in PCR inhibition. Wash the plates with 1–2 ml of

Table 5.4 Sensitivity of real-time PCR for detecting *R. solanacearum* bv 2 in potato

	Water	Potato extract		
	Direct PCR	Direct PCR	BIO-PCR	
CFU/ml	1 μl	1 μl	1 μl	10 μl
300,000	28[a]	32	27	24
30,000	30	35	31	28
3,000	33	38	34	31
300	–	–	36	33
30	–	–	–	36

[a]Numbers are cycle threshold values, defined as the cycle number at which fluorescence rises above the baseline

Table 5.5 Detecting *Xanthomonas citri* at Los Angeles International Airport by real-time PCR

Tissue	Origin	PCR (Ct)	Isolation
Leaf	Laos	32.8[a]	+[b]
Leaf	Cambodia	34.6	–
Fruit	India	–	–
Leaf	Thailand	31.2	+
Leaf	Cambodia	27.9	–
Leaf	Vietnam	28.7	–
Leaf	Thailand	33.2	–
Leaf	Vietnam	32.1	–

[a]Numbers are cycle threshold (Ct) values
[b]Isolation from leaf/fruit lesions on agar medium

water; use for direct PCR or extract DNA. Incubate the other three plates until the target colonies are large enough to identify for viable results. BIO-PCR can also be done using liquid media for enrichment (Song et al. 2004). A liquid protocol is much simpler and less labor intensive. BIO-PCR is designed to increase sensitivity and provide a viable result along with PCR. PCR assays have been developed for over 30 seed-borne pathogens (Fatmi et al. in press). Modifications of PCR, such as BIO-PCR, not only reduce inhibitors but greatly increase sensitivity by allowing the target bacterium to multiply. BIO-PCR expands our ability to detect low numbers of pathogens even in seeds contaminated with large numbers of saprophytes. The sensitivity of BIO-PCR is normally 100 times that of direct PCR without enrichment (Table 5.4, *Ralstonia*). Perhaps the most sensitive PCR protocol is to combine membrane filtration with BIO-PCR (Schaad et al. 2007). A major challenge for plant quarantine is dealing with PCR positive results but no culture for confirmation (Table 5.5, Los Angeles International Airport data).

Acknowledgements The senior author would like to thank the British Society of Plant Pathology and China Agriculture University Seed Health Centre for travel support to attend the tenth International Congress of Plant Pathology in Beijing, China.

References

Alvarez AM (2004) Integrated approaches for detection of plant pathogenic bacteria and diagnosis of bacterial diseases. Annu Rev Phytopathol 42:339–366

Bach H-J, Jessen I, Schloter M, Munch JC (2003) A TaqMan-PCR protocol for quantification and differentiation of the phytopathogenic *Clavibacter michiganensis* subspecies. J Microbiol Methods 52:85–91

Berg T, Tesoriero L, Hailstones DL (2006) A multiplex real-time PCR assay for detection of *Xanthomonas campestris* from brassicas. Lett Appl Microbiol 42:624–630

Bertolini E, Olmos A, Lopez MM, Cambra M (2003) Multiplex nested RT-PCR in a single closed tube for sensitive and simultaneous detection of four RNA viruses and *Pseudomonas savastanoi* pv. *savastanoi* in olive trees. Phytopathology 93:286–292

Cho MS, Lee JH, Her NH, Kim CK, Seol Y-J, Hahn JH, Baeg JH, Kim HG, Park DS (2012) A quantitative and direct PCR assay for the subspecies-specific detection of *Clavibacter michiganensis* subsp. *michiganensis* based on a ferredoxin reductase gene. J Microbiol 50:496–501

De Bellis P, Schena L, Cariddi C (2007) Real-time Scorpion- PCR detection and quantification of *Erwinia amylovora* on pear leaves and flowers. Eur J Plant Pathol 118:11–22

De León L, Rodríguez A, López MM, Siverio F (2008) Evaluation of the efficacy of immunomagnetic separation for the detection of *Clavibacter michiganensis* subsp. *michiganensis* in tomato seeds. J Appl Microbiol 104:776–786

Deuner CC, de Souza RM, Zacaroni AB, Figueira AR, Camera JN (2012) Sensitivity of the method of obtaining bacterial cells and PCR for detection of *Curtobacterium flaccumfaciens* pv. *flaccumfaciens* in bean seeds [Sensibilidade do método de obtenção das células bacterianas e da técnica de PCR para detecção de *Curtobacterium flaccumfaciens* pv *flaccumfaciens* em sementes de feijão]. Summa Phytopathol 38:48–53

Fanelli V, Cariddi C, Finetti-Sialer M (2007) Selective detection of *Pseudomonas syringae* pv. *tomato* using dot blot hybridization and real-time PCR. Plant Pathol 56:683–691

Fatmi M, Walcott R, Schaad NW (eds) Detection of bacteria in seed and other planting material, 2nd edn. APS Press, St. Paul (in press)

Guven K, Mutlu MB (2000) Development of immunomagnetic separation technique for isolation of *Pseudomonas syringae* pv. *phaseolicola*. Folia Microbiol 45:321–324

Hardoim PR, Hardoim CCP, van Overbeek LS, van Elsas JD (2012) Dynamics of seed-borne rice endophytes on early plant growth stages. PLoS One 7(2). Art. No. e30438

Hartung JS, Pruvost OP, Villemot I, Alvarez AM (1996) Rapid and sensitive colorimetric detection of *Xanthomonas axonopodis* pv. *citri* by immunocapture and nested-polymerase chain reaction assay. Phytopathology 86:95–101

He Y, Munkvold GP (2012) Comparison of extraction procedures for detection of *Xanthomonas axonopodis* pv. *phaseoli* in common bean seed. Plant Pathol 61:837–843

Holland PM, Abramson RD, Watson R, Gelfand DH (1991) Detection of specific polymerase chain reaction product by utilizing the 5′ to 3′ exonuclease activity of *Thermus aquaaticus* DNA polymerase. Proc Natl Acad Sci 88:7276–7280

Johnson KL, Walcott RR (2012) Progress towards a real-time PCR assay for the simultaneous detection of *Clavibacter michiganensis* subsp. *michiganensis* and Pepino mosaic virus in tomato seed. J Phytopathol 160:353–363

Kim BK, Cho MS, Kim MH, Choi HJ, Kang MJ, Shim HS, Ahn T-Y, Kim J, Park DS (2012) Rapid and specific detection of *Burkholderia glumae* in rice seed by real-time bio-PCR using species-specific primers based on an rhs family gene. Plant Dis 96:577–580

Koenraadt H, Van Betteray B, Germain R, Hiddink G, Jones JB, Oosterhof J, Rijlaarsdam A, Roorda P, Woudt B (2009) Development of specific primers for the molecular detection of bacterial spot of pepper and tomato. Acta Hortic 808:99–102

Maskos U, Southern EM (1992) Oligonucleotide hybridizations on glass supports: a novel linker for oligonucleotide synthesis and hybridization properties of oligonucleotides synthesized in situ. Nucleic Acids Res 20:1679–1784

Özdemir Z (2009) Development of a multiplex PCR assay for the simultaneous detection of *Clavibacter michiganensis* subsp. *michiganensis*, *Pseudomonas syringae* pv. *tomato* and *Xanthomonas axonopodis* pv. *vesicatoria* using pure cultures. J Plant Pathol 91:495–497

Palacio-Bielsa A, Cambra MA, Lopez MM (2009) Letter to the Editor. PCR detection and identification of plant-pathogenic bacteria: updated review of protocols (1989–2007). J Plant Pathol 91:249–297

Popović T, Balaz J, Nikolić Z, Starović M, Gavrilović V, Aleksić G, Vasić M, Zivković S (2010) Detection and identification of *Xanthomonas axonopodis* pv. *phaseoli* on bean seed collected in Serbia African. J Agric Res 5:2730–2736

Prosen D, Hatziloukas E, Schaad NW, Panopoulas NJ (1993) Specific detection of *Pseudomonas syringae* pv. *phaseolicola* DNA in bean seed by polymerase chain reaction-based amplification of a phaseolotoxin gene region. Phytopathology 83:965–970

Robène-Soustrade I, Legrand D, Gagnevin L, Chiroleu F, Laurent A, Pruvost O (2010) Multiplex nested PCR for detection of *Xanthomonas axonopodis* pv. *allii* from onion seeds. Appl Environ Microbiol 76:2697–2703

Saettler AW, Schaad NW, Roth DA (eds) (1989) Detection of bacteria in seed and other planting material. APS Press, St. Paul

Sakthivel N, Mortensen CN, Mathur SB (2001) Detection of *Xanthomonas oryzae* pv. oryzae in artificially inoculated and naturally infected rice seeds and plants by molecular techniques. Appl Microbiol Biotechnol 56:435–441

Schaad NW, Berthier-Schaad Y, Knorr D (2007) A high throughput membrane BIO-PCR technique for ultra sensitive detection of *Pseudomonas syringae* pv. *pheseolicola*. Plant Pathol 56:1–8

Schaad NW, Berthier-Schaad Y, Sechler A, Knorr D (1999) Detection of *Clavibacter michiganensis* subsp. *sepedonicus* in potato tubers by BIO-PCR and an automated real-time fluorescence detection system. Plant Dis 83:1095–1100

Schaad NW, Cheong SS, Tamaki S, Hatziloukas E, Panopoulas NJ (1995) A combined biological and enzymatic amplification (BIO-PCR) technique to detect *Pseudomonas syringae* pv. *phaseolicola* in bean seed extracts. Phytopathology 85:243–248

Schaad NW, Frederick RD, Shaw J, Schneider WL, Hickson R, Petrillo MD (2003) Advances in molecular-based diagnostics in meeting crop biosecurity and phytosanitary issues. Annu Rev Phytopathol 41:305–324

Song WY, Kim HM, Hwang CY, Schaad NW (2004) Detection of *Acidovorax avenae* spp. *avenae* in rice seed using BIO-PCR. J Plant Pathol 152:667–676

Umesha S, Chythra R, Kavitha R, Niranjana SR, Prakash HS, Mortensen CN (2010) Molecular detection of *Xanthomonas citri* subsp. *malvacearum* in cotton seeds. Res J Biotechnol 5:20–26

Veena MS, van Vuurde JW (2002) Indirect immunofluorescence colony staining method for detecting bacterial pathogens of tomato. J Microbiol Methods 49:11–17

Walcott RR, Gitaitis RD (2000) Detection of *Acidovorax avenae* subsp. *citrulli* in watermelon seed using immunomagnetic separation and the polymerase chain reaction. Plant Dis 84:470–474

Weller SA, Elphinstone JG, Smith NC, Boonham N, Stead DE (2000) Detection of *Ralstonia solanacearum* strains with a quantitative, multiplex, real-time, fluorogenic PCR (TaqMan) assay. Appl Environ Microbiol 66:2853–2858

Chapter 6
Improved Detection and Monitoring of Seed-Borne Fungal Plant Pathogens in Europe

Giovanni Vannacci, Sabrina Sarrocco, and Angelo Porta-Puglia

Abstract The main goal of seed pathology research and practice is the production and dissemination of high-quality, disease-free seed that maximizes potential crop productivity and value. Presently, the largest part of official seed health tests requires the growth of pathogens (direct methods) but molecular biology offers new tools to diagnose fungal pathogens in/on seeds (indirect methods), reducing the time required by direct methods and improving the output of seed health tests. However, molecular methods suffer some drawbacks and require a more difficult validation procedure than direct methods. This chapter aims at describing the historical context, and the development and implementation of methods for the detection and monitoring of seed-borne plant pathogenic fungi in Europe, with special emphasis on innovative methods.

Keywords Seed health test • Molecular methods • Multiplexing • Validation • Legislation • Quarantine • History • PCR • Multiplex PCR • Real time PCR • LAMP • DNA array

1 Introduction

Seed-borne pathogenic fungi affect directly and indirectly seed production, causing reduction in yields and seed quality. Through infected or infested seeds, diseases may enter and establish into new areas thereby causing quarantine concern and economic losses. Seed-borne inoculum may negatively affect germination in laboratory tests, and reduce to different extents emergence in the seedbed or in the field. Moreover, some fungal species, which may have no impact in the field, may reduce seed viability during storage.

G. Vannacci (✉) • S. Sarrocco
Dipartimento di Scienze Agrarie, Alimentari e Agro-ambientali, Università di Pisa, Pisa, Italy
e-mail: giovanni.vannacci@unipi.it; sabrina.sarrocco@unipi.it

A. Porta-Puglia
Istituto Sperimentale per la Patologia Vegetale (Presently CRA-PAV), Rome, Italy
e-mail: angelopp@alice.it

M.L. Gullino, G. Munkvold (eds.), *Global Perspectives on the Health of Seeds and Plant Propagation Material*, Plant Pathology in the 21st Century 6,
DOI 10.1007/978-94-017-9389-6_6

Although, in a globalised world, focusing on Europe may appear at first sight either vain or limited, the present status of seed pathology may be better understood, and the future prospects more easily forecast, if the geographical and related historical contexts within Europe are taken into account.

The progress of scientific studies on agriculture in the second part of the nineteenth century resulted also in increased interest for seed quality and led to the establishment of seed testing stations in several European countries. Seed pathology took its first steps as an independent discipline in Europe. Then, the newborn branch of science and its related technologies progressively disseminated to other continents and more and more countries contributed to its advancements (Munkvold 2009). The study of seed-borne plant pathogens, including fungi, was recognized as a special branch rather recently in the history of plant pathology and the name of 'Seed pathology' was probably first used in the '40s of the last century (Agarwal and Sinclair 1997). The rising branch of knowledge aimed firstly at integrating the existing seed-quality laboratory testing for purity, germination, vigour, etc. in use within the seed testing stations, by detecting potentially noxious micro-organisms present in, or associated with, seed. In the year 1919, L.C. Doyer was appointed as 'seed pathologist', in a special Division for studying the sanitary condition of the seed added to the Seed Testing Station at Wageningen, the Netherland, for the first time in the world. She was to become, within the International Seed Testing Association (ISTA, founded in Cambridge, UK, in 1924), the first president of the Committee for the determination of seed-borne diseases, presently the Seed Health Committee (SHC), followed in this commitment by W.F. Croiser (1949–1953), A.J. Skolko (1953–1956) and P. Neergaard (1956–1974).

In 1967, the Danish Ministry of Foreign Affairs founded the Danish Government Institute of Seed Pathology for Developing Countries (DGISP) and P. Neergaard was its first director (1967–1980), followed by S.B. Mathur (1980–2003). DGSIP promoted the scientific knowledge of seed pathology in Denmark and in the world and trained hundreds of specialists, particularly from developing countries.

The contribution of European authors to the advancement of seed pathology in general, and to detection and monitoring of seed-borne pathogens, was outstanding. To support this fact we just wish to mention here three seminal publications of the past: (i) the detailed manual by Doyer (1938), which was used for decades in the laboratories all over the world, (ii) the Annotated list of seed-borne diseases, by M. Noble (Scotland), J. de Tempe (the Netherland) and P. Neergaard (Denmark), produced under the auspices of ISTA and published in 1958 by the Commonwealth Mycological Institute (Noble et al. 1958), listing approximately 900 plant diseases that may be spread by seed-borne organisms. A second edition followed (Noble and Richardson 1968). The third edition was authored by Richardson (1979) and was successively updated in supplements, (iii) the textbook by Neergaard (1979) which was worldwide a basic tool for generations of students and scholars alike.

The discovery and subsequent applications of nucleic-acid techniques in plant pathology promoted innovation in the detection of seed-borne pathogens and in the study of the epidemiology of seed-transmitted diseases. Nevertheless the impact of the recent tools on the standard methods used in seed health for quarantine and quality routine testing is still limited, particularly as concerns fungi. We will describe some of the drawbacks and illustrate the perspectives for future advancements which can be expected on the basis of newly developed diagnostic technologies.

2 Detection of Seed-Borne Fungi

The main goal of seed pathology research and practice is the production and dissemination of high-quality, disease-free seeds that maximize potential crop productivity and value. Since infected seeds are the most important carrier of pathogens for trans-regional and long distance dissemination, the detection and identification of plant pathogens, a key factor in seed pathology together with elucidation of epidemiology, is basic to prevent the introduction and spread of new pathogens in areas where they are not present (Munkvold 2009).

Detection of pathogenic fungi, generally in plant material and particularly in seeds, can be difficult, due to several factors: the pathogen often is present at low infection levels; samples may be affected by related species of pathogens which may cause similar symptoms; some pathogens cannot easily be cultured in vitro; and most methods are not able to accurately quantify the pathogen. Ideally, seed assays should be sensitive, specific, rapid, robust, inexpensive and simple to implement and interpret (Walcott 2003). The development of more versatile, robust and cost effective systems, together with greater sensitivity and specificity, elevated throughput and detection of multiple pathogens, is increasingly necessary to improve disease control decision-making (Majumder et al. 2013). Seed assays have been developed based on different techniques including visual examination, selective media, seedling grow-out, serological techniques and nucleic-acid techniques. Each technique has advantages and shortcomings and most of them are far from being ideal.

2.1 Traditional Methods

Some classic diagnostic methods have been developed since long ago and applied to routine seed testing (Neergaard 1979; De Tempe and Binnerts 1979; Cappelli and Covarelli 2005). The most common methods are briefly discussed below.

Examination of Ungerminated Seed Signs of pathogens (sclerotia, mycelial mats) and symptoms of disease (discolouration, pigmentation, etc.) are detected by visual inspection of the dry seed (*dry inspection*) by the naked eye or with the help of optical devices (hand lenses, stereomicroscopes). Liquid drops may be put on seeds to see fruiting bodies of fungi or to facilitate the oozing of spores; suspensions may be obtained by washing seed in water or other liquids to separate spores, conidia, or other specific structures from the seed. Fungal structures may be separated by centrifugation, or collected on the surface of filters, and may be identified and counted with the help of microscopy (*washing tests*). These methods provide quick information and allow easy detections, e.g., of sclerotia (*Claviceps* spp.). However, only a few species of pathogens are detectable and these methods do not permit distinction between live and dead organisms. Therefore these techniques are mostly considered as preliminary and complementary to other methods.

Incubation Methods Seeds are incubated either on water-soaked blotter/filter paper (*blotter tests*) or on agar media (*agar plate tests*) under controlled temperature and light conditions (20 ± 2 °C, NUV wave length peak around 360 nm alternated with dark, 12 h photoperiod, being used for a large range of fungal species). After appropriate incubation each seed is examined microscopically and/or by the naked eye to detect fungal species on the basis of their morphology. Different agar media may be used, some being selective or semi-selective. Various kind of pretreatments may be applied (e.g. surface disinfestation). A variant for blotter test includes treatment of incubated seeds, once water from the substrate has saturated their tissues, at −20 °C for 24 h (*deep-freezing blotter method*) or the addition of herbicides (2.4 D) to kill the embryo. In these ways seed germination in the plates is inhibited and development of fungal colonies and their observation are facilitated. Incubation methods are efficient and widely used for the detection of many seed-borne pathogens, nevertheless they may be unsuitable for fungicide-treated seed or require previous removal of fungicides when applicable. If the seed is incubated without germination-inhibiting treatments, symptoms may show up on shoots or rootlets, thus providing indications on the virulence of the inoculum present in the seed sample.

Embryo Test When infection is localized in the embryo, as it is the case with *Ustilago nuda* in barley, the seed may be softened by chemical treatment and embryos separated and rescued for examination under the stereomicroscope for detection of the intra-embryonal fungal mycelium. Variants have been proposed to avoid the use of the toxic phenol and to reduce the number of embryos to be examined per seed sample (Khanzada et al. 1989; Cappelli et al. 1993).

Serological Methods Largely used for viruses and bacteria, these techniques also have been applied for the detection of some fungal species. In seed-health testing, an immunoblot method, proposed by Hill et al. (2002), is included in the current ISTA rules to detect *Neotyphodium* spp. in *Festuca* spp. and *Lolium* spp.

Growing-on Methods Seed may be sown on different substrates (sand, brickstone, etc.) in trays, or in testing tubes on agar (one seed per tube), and seedlings are grown under conditions suitable for the appearance of symptoms of the disease or sign of the pathogen. These methods may take longer time than other methods, nevertheless they provide reliable information about the virulence of the pathogens and may, to some extent, be predictive of the field or seedbed performance of a seed lot.

2.2 Innovative Methods

Polymerase Chain Reaction (PCR) Molecular-based methods began to develop after the introduction of PCR in the mid 1980s and a rapid development of genomic techniques lays the foundations for improving plant pathogen detection and identification. More and more diagnostic laboratories and inspection agencies use molecular methods for the identification and detection of seed-borne pathogens on different crops. In general, these methods are faster, more specific, more sensitive and more accurate, and can be performed and interpreted by personnel with no specialized taxonomical expertise. These techniques allow the detection and identification also of non-culturable microorganisms and, due to their high degree of specificity, can distinguish closely related organisms at different taxonomic levels (Capote et al. 2012) and have sensitivity to detect low infection rates (Konstantinova et al. 2002).

The internal transcribed spacer (ITS1, 5,8S sequence and ITS2) region of the rDNA is widely used to design species-specific primers to unequivocally detect the presence of pathogenic fungal species in plant tissues and to identify fungal species in pure culture. Otherwise, other species-specific primers have been designed and used for PCR detection of seed-borne fungi. PCR protocols have been developed for the detection of *Alternaria* spp. in cruciferous seed (Iacomi-Vasilescu et al. 2002), carrot (Konstantinova et al. 2002), linseed (McKay et al. 1999) and sunflower (Udayashankar et al. 2012), *Phoma valerianellae* in lamb's lettuce seeds (Pellegrino et al. 2010) and *Peronospora belbahrii* in basil seeds (Djalali Farahani-Kofoet et al. 2012). Many papers have been published reporting the use of DNA-based methods for the detection of a number of bunt fungi as *Tilletia caries* (syn *T. tritici*) and *Tilletia controversa*, causing common and dwarf bunt of wheat, respectively (Josefsen and Christiansen 2002; Kochanová et al. 2004; Eibel et al. 2005; Kellerer et al. 2006; Roberts et al. 2007; Zouhar et al. 2010).

Multiplex PCR Multiplex PCR allows the simultaneous and sensitive detection of different DNA and RNA targets in a single reaction, helping in reducing the number of tests required (Majumder et al. 2013). Multiplex PCR found successful application particularly for *Fusarium* species within the Fusarium wheat head scab complex, allowing the detection by a single amplification experiment of the major species involved in the disease (Waalwijk et al. 2003). This technique has been established also for the identification of different chemotypes among *Fusarium graminearum*, *Fusarium culmorum* and *Fusarium cerealis* contaminating wheat kernels. The rapidity of the assay could allow a fast screening of large numbers of samples for potentially toxigenic fungal species/populations with a possible application in the agro-food industry (Quarta et al. 2006).

Loop-Mediated Isothermal Amplification (LAMP) A quite new nucleic amplification technique, loop-mediated isothermal amplification (LAMP) has been described as an easy and rapid diagnostic tool for the early detection of microbes (Parida et al. 2008; Vincelli and Tisserat 2008) with levels of sensitivity and selectivity equivalent to PCR without the requirement of sophisticated thermocyclers (Kubota et al. 2007). This technique is a simple and rapid procedure for the specific detection of genomic DNA using a set of six oligonucleotide primers with eight binding sites hybridizing specifically to diverse areas of a target gene and a thermophilic DNA polymerase from *Geobacillus stearothermophilus* for DNA amplification (Notomi et al. 2000). Only a few application of LAMP for the detection of seed-borne pathogens have been described in Europe to date, such as the development and application of LAMP-based assay for the detection and identification of *F. graminearum* in contaminated samples of wheat and barley seeds (Niessen and Vogel 2010; Abd-Elsalam et al. 2011) to improve quality management in the cereal industry by making analysis easier and more cost effective.

Real-Time PCR In the early 1990s, with the development of the TaqMan® chemistry by Applied Biosystems (Foster City, CA, USA) (Holland et al. 1991), a real breakthrough in reliable, simple amplicon detection came out. Initially the costs for real-time PCR machines and reagents limited its use to research activities rather than for diagnostic purposes (Cullen et al. 2002; McCartney et al. 2003; Gachon et al. 2004), but these costs have been reduced over time. Over the past decade, real-time PCR technology (qPCR) has been developed in lieu of conventional PCR assays. Factors that make real-time PCR a better established diagnostic technique over PCR are the robustness of amplification, greater sensitivity, the lack of post-PCR manipulations and the level of skills required for by operators, that is similar to that of ELISA (Boonham et al. 2008) and the quantitative results, expressed as the amount of target DNA present in a sample.

Because no post PCR manipulation is required (particularly the need for gel electrophoresis is removed), real-time PCR is less laborious than conventional PCR and it is therefore suitable for automation and high throughput testing (Gil-Serna et al. 2009). High throughput real-time PCR machines able to process DNA samples continuously over a 24 h period, in 384-well plates, will further facilitate its use for rapid diagnosis of a large numbers of seed or tissue samples (McCartney et al. 2003; Guillemette et al. 2004).

Reliability of qPCR analysis is of paramount importance, due to economic and environmental consequences of destruction or rejection of false-positive samples and introduction of the pathogens in disease-free areas due to false-negative samples (Ioos et al. 2012). By using this procedure it is possible not only to detect the presence of a target pathogen but also accurately quantify the amount present in plant material (Mumford et al. 2006). Very often, seed infections occur at very low frequencies; a procedure as sensitive as real-time PCR allows reducing the risk of planting seed lots with low frequencies of infection. This technique is useful to remove some of the ambiguities due to small sample size and low sensitivity (Chen et al. 2013).

Real-time PCR-based methods have been applied for the identification and detection of *Pyrenophora* species in barley seeds allowing the quantification of pathogen DNA extracted from infected seed to the picogram level by using the fluorescent reporter dye SYBR Green and completing the test in 8 h, compared to 7 days for the traditional agar plate test (Bates et al. 2001). Bates and Taylor (2001) described the development of a Scorpion ARMS primer for the specific detection and quantification of *Pyrenophora teres*. Combination between Scorpion and amplification refractory mutation system (ARMS) enables single base mutations to be detected, allowing the discrimination of closely related species. By using these sensitive and specific fluorescent probes it was possible to specifically detect *P. teres* in a single reaction, whilst previously two reactions were required to discriminate *P. teres* from *P. graminea* on barley seeds.

Papers reporting the use of real-time PCR as a diagnostic tool for plant pathogens, including fungi, are becoming more numerous. A TaqMan® real–time PCR method was developed to monitor and quantify the dynamics of individual species within the complex involved in Fusarium head blight of winter wheat (Waalwijk et al. 2004) furnishing a versatile tool that has been applied in a comparison of different genotypes and that could be also applied to other disease management systems. A quantitative real-time PCR assay using TaqMan chemistry has also been developed to quantify the level of *Tilletia* spp. contamination in wheat-seed lots in UK, allowing an increase in test throughput and providing the sensitivity required for an advisory threshold of one spore per seed (McNeil et al. 2004). TaqMan® probe was also used in a real-time assay to detect the causal agent of wilt and crown rot of basil from infected plants and seed in Italy in order to diminish testing time, to identify both internally and externally infected seed with a high sensitivity and to create the conditions for future automation (Pasquali et al. 2006).

A real-time PCR assay for the early detection of *Fusarium fujikuroi* at very low levels in rice seeds was developed in Italy, providing a potentially very useful tool to reduce the risk of commercializing infected seed lots or in taking decisions on appropriate management strategies in order to prevent the spread of bakanae disease in the field (Amatulli et al. 2012).

As for conventional PCR, real-time PCR still can suffer from interference from inhibitory compounds in seed extracts. To overcome this problem, combination of real-time PCR protocols with procedures to separate pathogen DNA from inhibitory compounds and non-target DNA, such as magnetic capture hybridization (MCH), have been developed (Munkvold 2009). MCH was successfully demonstrated with *Botrytis aclada* from onion seed (Walcott et al. 2004), increasing sensitivity at least tenfold compared to direct real-time PCR (Walcott et al. 2008).

A combination of an enrichment procedure with qPCR facilitated sensitive detection of *Fusarium circinatum*, the causal agent of pitch canker disease on numerous *Pinus* spp., in pine seed. This aggressive fungus may infect pine seed cryptically and, therefore, can easily be spread long distances by the seed trade. The dual-labelled probe-based-real-time PCR test developed by Ioos et al. (2009) has proved to be highly specific because it did not cross-react with DNA from phylogenetically close species and is significantly more sensitive than conventional PCR, enabling the detection of the pathogen in samples artificially contaminated with less than 1/1,000 infected seeds as well as in naturally infected samples.

A new real-time PCR test targeting *Plasmopara halstedii* was developed, optimizing a duplex real-time PCR tool to target the pathogen within sunflower seeds and maximizing the analytical sensitivity without compromising the specificity.

The authors also provided a fully optimized DNA extraction protocol, a step of paramount importance because the pathogen cannot be sub-cultured and biologically amplified, to improve sensitivity of the test and to eliminate chemical compounds present in the seed, especially the hull, that may inhibit the PCR reaction. Finally, to ensure reliability of the results, a set of controls was used systematically during the duplex real-time PCR, including a plant-specific probe (Ioos et al. 2012).

Other Multiplex Techniques Several plant pathogens can be seed transmitted in the same host plant. Each of these pathogens requires a specific seed health test, so a single seed lot may need to be evaluated by a number of seed health tests equal to the number of pathogens transmitted by seeds. Usually, at harvest, many seed lots need to be certified for the absence of specific pathogens in a quite short time and this can be achieved by parallel, or high throughput, plant pathogen detection techniques. Several excellent papers have been recently published that review this subject in detail (Tsui et al. 2011; Capote et al. 2012; De Boer and Lopez 2012) but little information is available specifically for detection of fungal pathogens in/on seeds.

Multiplex detection techniques offer the possibility to reduce the costs and the time required to define the health status of a seed lot by single-pathogen detection techniques. Molecular biology offers several tools, but at the moment the most exploited are based on real-time PCR (Waalwijk et al. 2004). The limited number of available dyes and the need to limit the number of fluorescent signals in a single reaction allow the detection of only a few pathogens in a single test. Multiplex real time PCR can be incorporated in seed health tests targeted to detect both RNA- and DNA-based pathogens (Ling et al. 2011) or to detect specific mycotoxin-producing genotypes (Kulik et al. 2011).

To improve multiplex plant pathogen detection, DNA array technology has been developed (Zhang et al. 2007). This technology is based on the visualization of hybridization reactions between nucleic acids in a sample with unique nucleic acid sequences of known pathogens immobilized on a solid surface. Microarrays can detect up to thousands of unique sequences, and the range of organisms detected can be expanded to include not only all the different pathogens which can affect a host plant but also specific plant features, such as the variety or the presence of GM traits (Germini et al. 2005). However, production of hybridisation signals can be highly variable and extremely sensitive to minor technical differences (Carmichael 2012). Microarray format can be planar, with probes bind to the surface of a two dimensional surface, such as glass slides, but to achieve a higher working capacity probes can be linked to the surface of beads suspended in fluid (Peters et al. 2007; http://www.luminexcorp.com). Microarray can be designed to detect a group of pathogens (for viruses, see http://www.bio-chip.co.uk/) or all the pathogens affecting a specific crop or group of crops (for example see http://www.dnamultiscan.com). But the pace of technology is running fast, and new tools are made available by industries that can be adopted to detect seed borne pathogens. One possible alternative is represented by padlock probes (PLP) (Szemes et al. 2005). These probes (recently reviewed by Tsui et al. 2013) comprise two target- complementary sequence regions at both ends for hybridization to specific DNA sequences, as well as a non target-complementary segment. Upon hybridization to the target, the two ends are brought into contact, allowing PLP circularization by ligation which is followed by universal amplification and microarray detection. PLPs (called PRI-probes) have been combined (van Doorn et al. 2007) with the OpenArrays™, which can accommodate up to 3,072 33 nl PCR amplifications, allowing high-throughput real-time quantification and high multiplexing ability.

But in the near future a different approach for seed health testing could be envisaged. A molecular profiling of the whole microbial community associated with seeds can be defined by molecular techniques such as denaturing gradient gel electrophoresis (DGGE) (Dent et al. 2004), but massively parallel next-generation sequencing (NGS) techniques coupled with metagenomics analysis and further fostered by a robust barcoding system (http://www.boldsystems.org/) has opened the possibility to decipher complex communities by sequencing target genes or sequences. The huge amount of sequence data generated by NGS (Tucker et al. 2009) represents any and all organisms present in the sample and can be exploited to detect plant pathogens (Adams et al. 2009) and to analyse fungal communities (Lindahl et al. 2013; Bokulich and Mills 2013), but requires a heavy computational effort to be analyzed. Attempts to reduce the time for bioinformatical work (Stobbe et al. 2013; White et al. 2013) have been made, but cost and the required technical skills make it, at the moment, unaffordable for routine seed analysis laboratories.

3 Every Rose has Its Thorns

Drawbacks of Molecular Techniques Molecular seed health testing methods suffer from some particular drawbacks. Positive detection of pathogens in most molecular diagnostic protocols is based on detection of pathogen-specific DNA sequences. This can result in false positive results, due to the presence of non-viable pathogen cells. Nucleic acid extraction from seeds can co-extract PCR-inhibiting compounds, and this can result in false negative tests. Sensitivity (proportion of true positive that test positive) and specificity (proportion of true negative that test negative) have to be evaluated during validation procedures, but their relative importance depends upon the intended use of the results of the test. Quarantine pathogens will require highly sensitive tests, accepting the risk to have false positives (Ioos et al. 2012) while detection of quality pathogens can rely on highly specific tests.

Quantification of pathogens in seeds is ambiguous. Molecular methods based on qPCR express the amount of the pathogen inoculum as the amount of target DNA present in a sample and a sample is usually made up of many (up to thousands) seeds. No information is provided about the number of infected seeds present in the seed lot, nor about the localization of the pathogen inoculum in/on the seeds. These two last pieces of information are of epidemiological value, as the number of infected seeds represents the number of possibly diseased plants (primary inoculum) randomly distributed in a field, and the inoculum localization affects the seed- to- plant transmission rate.

Finally, the implementation of molecular diagnostic methods brings a different kind of problem. Seed (plant) health testing has moved from plant pathology laboratories to biochemical laboratories where individual growers can request guidance about the health status of their seeds/plants. But a nice musical metaphor of Woese (2004) makes clear the inherent risk of such a drift: "molecular biology could read notes in the score, but it couldn't hear the music"; as to say that the output of a seed health test, for quality pathogens, needs to be evaluated by a plant pathologist that has the required competencies to properly interpret the results.

Molecular Techniques Validation Methods used for detection of seed-borne fungi in current official seed-health testing for quality are based on validated protocols included in the ISTA rules. The general principles and objectives for seed health testing, the lines to be adopted for validation of seed health methods, and how results must be reported are detailed in Chap. 7 of the *International Rules for Seed Testing*. Annex to Chap. 7 describes the validated methods. The strict requirements of standardization and ring-testing for validation to be fulfilled before a method is adopted results in a limited number of official protocols available for use. The current Annex 7 (ISTA 2014) lists methods for the detection of only 20 host-fungal pathogen couples.

In addition to ISTA, other organizations elaborate standardized seed health test methods for use in international trade, e.g. the International Seed Health Initiative (ISHI), and the National Seed Health System (NSHS) in the USA. A detailed description of organizations involved in seed analysis at the international level has been published (OECD 2012).

An in depth discussion of validation protocols is out of the scope of the present chapter, therefore we should remember that validation is the procedure through which a laboratory-developed seed health test method can be adopted by diagnostic laboratories to issue official certificates. Validation is the "confirmation, through the provision of objective evidence, that the requirements for a specific intended use or application have been fulfilled" (ISO 9000:2005; https://www.iso.org/obp/ui/#iso:std:iso:9000:ed-3:v1:en). In Europe few research laboratories and Institutions are involved in validation procedures. The International Seed testing Association, based on the work of the Seed Health Committee, decided, in 2002, to establish the Method Validation Working Group to stress that method validation should apply to all seed quality testing, not just tests for seed health (ISTA 2007). No tests can be included in the International Rules for Seed Testing unless they have passed a validation procedure. At the moment only three molecular methods have been approved by ISTA, but are all directed to detect seed transmitted bacteria. As far as regulated pests are concerned, EPPO adopts validated diagnostic protocols to detect fungal pathogens in seeds (EPPO 2007; EPPO 2008; EPPO 2009). In general, EPPO diagnostic protocols are used by official laboratories to detect and identify pests of potential phytosanitary concern in the EPPO region and provide requirements for reliable diagnosis. In cases where morphological tests can be reliable but appropriate molecular tests have been developed, the latter are presented as alternative or supplementary. Test are selected for sensitivity, specificity and reproducibility. Other factors such as easiness of use, availability of equipment and expertise required are taken into account. EPPO recommends that NPPOs use these criteria in order to determine the test, or combination of tests, appropriate for the circumstances (EPPO 2010).

The large gap existing between the number of published molecular seed health test methods and those included in the lists of official methods emphasizes the difficulties of the validation process. Validation is a long and costly procedure that, as a rule, involves many laboratories and should furnish performance criteria such as analytical sensitivity, analytical specificity, reproducibility and repeatability (Ioos et al. 2013; Ioos and van den Boogert 2012). In addition to the sources of variability that affect "classical" seed health tests, technical parameters specific to molecular methods can affect results. For example, the protocol for nucleic acid extraction from matrices, the reagents used in all the different steps and the equipment, and its calibration, used to amplify target sequences, can all affect results. These features, including the brand of commercially available reagents and equipment, need to be evaluated during the validation procedure. The continuous improvement of molecular techniques and the strong reliance on fine chemicals and highly technological equipment produced by the industry pose the risk that once a molecular method has been validated, in a short time it becomes obsolete or it requires a new round of validation due to the differences arising in the meantime in reagents and equipments quality. In other words, validation should be considered a continuous process. One more feature needs to be evaluated, the ruggedness (robustness) of the method. All diagnostics test can be affected by the physical environment or by operational conditions. Factors affecting direct "classical" tests are well known (Neergaard 1979), but molecular procedures are far more complicated and even with well defined protocols, minor variation of the many factors involved cannot be avoided. Validation procedures must take care of this aspect and validated protocols have to be robust enough to withstand such variability (Ioos et al. 2012).

4 Quarantine Legislation and Monitoring

Seed borne species of fungi of quarantine concern, as well as any other pest, are dealt with within a complex technical and legislative international system based upon:

(i) the International plant Protection Convention (IPPC) (https://www.ippc.int/) which come into force in 1952 and currently includes 181 Member countries. IPPC has the goal to protect the world's cultivated and natural plant resources from the spread and introduction of plant pests while minimising interference with the international movement of goods and people. The IPPC Secretariat is hosted and provided by The Food and Agriculture Organisation of the United

Nations (FAO), Rome, Italy. The Convention is governed by the Commission on Phytosanitary Measures (CPM);

(ii) The Agreement on the Application of Sanitary and Phytosanitary Measures (the "SPS Agreement" (http://www.wto.org/english/tratop_e/sps_e/spsagr_e.htm), concerning the application of food safety and animal and plant health regulations, which entered into force with the establishment of the World Trade Organization (WTO) in 1995.

As a result of the above, phytosanitary measures against any harmful organism adopted by a country have to be commensurate with the risk posed by the organism and science-based, transparent justifications are required. The IPPC International Standard for Phytosanitary Measures on Pest Risk Analysis (ISPM) No. 11 (FAO 2013) gives detailed guidance for the conduct of a pest risk analysis (PRA) to determine if an organism is a quarantine pest, and describes how to select risk management options.

IPPC provides a framework for international cooperation and harmonisation between contracting parties. These aims are achieved through the action of the National Plant Protection Organisations (NPPOs) which are the official government bodies in charge of the functions specified by the IPPC for each country, and of the Regional Plant Protection Organizations (RPPOs), which are the coordinating bodies at a regional level for the activities related to the IPPC objectives. RPPO recommendations are considered as Regional Standards in the sense of the IPPC. One of the RPPOs priorities is to prevent the introduction of dangerous pests from other parts of the world, and to limit their spread within the region in case of introduction. The risk of introducing pests to new areas has greatly increased in recent years with the expansion of trade. Seed is a potentially powerful means of geographical diffusion of many plant pathogens. In addition to commercial cultivars and hybrids for marketing, seed movement includes germplasm for research and development, experimental lines for breeding purposes and basic seed for multiplication.

The European and Mediterranean Plant Protection Organisation (EPPO), Paris, France, is the intergovernmental RPPO for the geographical region which includes Europe. EPPO makes recommendations concerning phytosanitary issues to the Member Governments (51 at present). In order to fulfil its RPPO commitments, EPPO collects information on pests, and produces standards through a complex approval procedure to assure their international acceptance. One of the EPPO standards lists the organisms recommended for regulation as quarantine pests. Pests are grouped into lists: A1 (not present in the region), and A2 (present but not widely distributed there and being officially controlled) (EPPO 2013). This standard is periodically reviewed and amended on the basis of scientific documentation and expert judgment. Each country need not include each listed pest into their national legislation: a PRA process for each pest will identify the endangered areas within the EPPO region.

Diagnostic protocols are also included among the EPPO standards. One example, concerning a seed borne pathogen, is the protocol PM 7/29 (2) for *Tilletia indica* Mitra (EPPO 2007, at the moment under revision: see https://www.eppo.int/STANDARDS/council2013.htm). This protocol starts with a washing test to identify the pathogen using morphological characters. If a limited number of spores (less than 10) is detected, making a morphological identification unreliable, then molecular confirmation by PCR using species-specific primers or combined as appropriate with restriction enzyme analysis is recommended.

As concerns quarantine in the European Union (EU), the Member States (MS) through their respective NPPOs, operate within a framework which aims at harmonising the national legislations in agreement with the EU legislation. The EU phytosanitary system is an open regime: movements of plants and plant products into and within EU are allowed on condition that specific restrictions and requirements are respected. The basic reference document on quarantine regulation is the Council Directive 2000/29/EC of 8 May 2000 on protective measures against the introduction into the Community of organisms harmful to plants or plant products and against their spread within the Community (consolidated version: http://eur-lex.europa.eu/LexUriServ/LexUriServ.do?uri=CONSLEG:2000L0029:20130411:EN:PDF). The directive includes 29 articles and nine annexes. The annexes of major interest for seeds are:

Annex I, Part A – Harmful organisms whose introduction into, and spread within, all Member States shall be banned – Section I – Harmful organisms not known to occur on any part of the Community and relevant for the entire Community, which includes *T. indica*;

Annex II – Part A – Harmful organisms whose introduction into, and spread within, all Member States shall be banned if they are present on certain plants or plant products. In *Section II – Harmful organisms known to occur in the Community and relevant for the entire Community, Plasmopara halstedii* (Farlow) Berl. and de Toni on seeds of *Helianthus annuus* L. is listed.

Annex II – Part B – Harmful organisms whose introduction into, and spread within, certain protected zones shall be banned if they are present on certain plants or plant products. Here *Glomerella gossypii* Edgerton on seeds and fruits (bolls) of *Gossypium* spp. is included, for the protected zone of Greece;

Annex IV – Part A – Special requirements which must be laid down by all Member States for the introduction and movement of plants, plant products and other objects into and within Member States. In *Section I*, concerning items originating from outside the Community, special requirements are detailed for (i) seeds of *H. annuus* concerning *P. halstedii* (i.e. seed must be produced in areas free from the pathogen, or appropriately treated against it if not produced on cultivars resistant to all its known races) and (ii) seed of *Triticum*, *Secale* and × *Triticosecale* originating from specific countries where *T. indica* is known to occur (statement of production in a pest-free area). Special requirements concerns also grains of *Triticum*, *Secale* and × *Triticosecale* from the mentioned origin: either grains originate from pest-free areas or absence of symptoms on plants at the place of production during their last complete cycle of vegetation is required and representative samples of the grains, taken both at the time of harvest and before shipment, are tested and found free from *T. indica*). In *Section II*, which deals with items originating in the Community, seeds of *H. annuus* are subject to the special requirements mentioned above for this species. *Part B* regards protected zones: special requirements are mentioned for seeds of *Gossypium* spp. about *G. gossypii* (Greece);

Annex V (inspections, plant passport, etc.) and *Annex VII* (phytosanitary certificate) should also be mentioned here.

Moreover plant health aspects of seed are included in other EU directives: Council Directive 66/401/EEC on the marketing of fodder plant seed, Council Directive 66/402/EEC on the marketing of cereal seed, etc., i.e. the so called legislation on the marketing of seed and plant propagating material (SPPM). Consolidated versions (i.e. which take into account all the amendment to the original directives) of the EU legislation on seed are available (http://eur-lex.europa.eu/collection/eu-law/consleg.html).

In 2010, an evaluation of the plant health regime in EU identified the main problems related with the present EU legislation. Among them, the insufficient focus on prevention in relation to increased imports of high-risk commodities, the need for prioritising pests at the EU level across all the Member States, and the need for better instruments for controlling the presence and natural spread of pests in case they reach the EU territory. Moreover, a need for upgrading the instruments concerning intra-EU movements (plant passports and protected zones) was pointed out. This analysis has resulted in suggestions to amend appropriately the current EU legislation. Therefore the European Commission, in May 2013, published a Proposal for a Regulation of the European Parliament and of the Council on protective measures against pests of plants. The proposal is aimed at overcoming flaws and putting in place a robust, transparent and sustainable regulatory framework for the EU Member States. The proposed Regulation is intended to replace Directive 2000/29/EC. The proposal reinforces the synergies with the plant reproductive material regime, while removing avoidable duplications. Among other changes, the pests currently regulated under SPPM directives are replaced under the proposed plant health Regulation and the status of widespread quarantine pests is changed into quality pests (http://ec.europa.eu/food/plant/plant_health_biosafety/rules/index_en.htm). The document has been submitted to the European Parliament and Council for co-decision. The procedure will take time; therefore no basic change in the current legislation is expected within the next few years.

National Plant Protection Organizations (NPPOs), acting within the legislative framework of their respective countries, inspect imported regulated plant materials at the official entry points, or elsewhere when applicable, take the necessary actions when quarantine pest are intercepted (reporting, eradication, etc.); they survey the territory to monitor plant health. Relevant findings by NPPOs of the EPPO member countries are published on the EPPO Reporting Service (http://www.eppo.int/PUBLICATIONS/reporting/reporting_service.htm), which includes interceptions, first reports and outbreaks of pests, new host plants, reappearance of pests considered no longer to occur in an area, denial of previous records, detection and identification methods, and other events of phytosanitary concern. Additionally, limited to EU countries and Switzerland, interceptions of non-compliant consignments are also reported electronically, via a direct web-link, on EUROPHYT, a web-based network and database connecting Plant Health Authorities of the EU Member States and Switzerland, the European Food Safety Authority (EFSA), Parma, Italy, and the Directorate General for Health and Consumers of the European Commission (http://ec.europa.eu/food/plant/plant_health_biosafety/europhyt/network_en.htm).

The recent interceptions of seed borne quarantine pests on seed mostly concern insects, bacteria and viruses. E.g., in the 12 months between 1st October 2012 and 30th September 2013 we have found no record of interception of seed borne quarantine fungi neither in the EPPO Reporting Service nor in EUROPHYT, while several records concern other organisms. Among other factors, seed certification schemes that include seed treatments, field inspections and field treatments may have largely contributed to improve the health condition of seed. The relatively easier detection of symptoms of fungal diseases compared to diseases of other aetiology during the seed multiplication process, and the availability of highly efficient fungicides, may have facilitated the efficiency of certification schemes in preventing seed borne infections and contaminations by fungi.

Innovative seed health testing methods attract attention all over the world. Innovation is, currently, mainly based on molecular tools. But molecular based diagnostics in Plant Pathology is today in a box: it heralds significant technological improvements but has been able to generate only few officially approved seed health testing methods. Anyway, traditional methods have shown their limits. Technology evolution, and signals coming from other research areas, suggest that molecular methods are on the right track and an increasing number of them may reasonably be expected to appear among officially validated methods. Then, adequate efforts and resources need to be deployed to ensure their integration among the tools available to implement an Integrated Pest Management strategy, as required, in Europe, by the Directive 2009/128/EC of the European Parliament establishing a framework to achieve the sustainable use of pesticides. But technical adjustments are only one side of the coin, the other being relaying on the expertise of plant pathologists to draw correct inferences from a seed health test output.

Acknowledgements GV wish to thank Valérie Grimault and Renaud Ioos for the helpful suggestions.

References

Abd-Elsalam K, Bahkali A, Moslem M, Osama E, Amin OE, Niessen L (2011) An optimized protocol for DNA extraction from wheat seeds and loop-mediated isothermal amplification (LAMP) to detect *Fusarium graminearum* contamination of wheat grain. Int J Mol Sci 12:3459–3472

Adams IP, Glover RH, Monger WA (2009) Next-generation sequencing and metagenomic analysis: a universal diagnostic tool in plant virology. Mol Plant Pathol 10:537–545

Agarwal VK, Sinclair JB (1997) Principles of seed pathology, 2nd edn. CRC Press, Boca Raton, p 539

Amatulli MT, Spadaro D, Gullino ML, Garibaldi A (2012) Conventional and real-time PCR for the identification of *Fusarium fujikuroi* and *Fusarium proliferatum* from diseased rice tissues and seeds. Plant Pathol 134:401–408

Bates JA, Taylor EJA (2001) Scorpion ARMS primers for SNP real-time PCR detection and quantification of *Pyrenophora teres*. Mol Plant Pathol 2:275–280

Bates JA, Taylor EJA, Kenyon DM, Thomas JE (2001) The application of real-time PCR to the identification, detection and quantification of *Pyrenophora* species in barley seed. Mol Plant Pathol 2:49–57

Bokulich NA, Mills DA (2013) Improved selection of internal transcribed spacer-specific primers enables quantitative, ultra-high-throughput profiling of fungal communities. Appl Environ Microbiol 79:2519–2526

Boonham N, Glover R, Tomlinson J, Mumford R (2008) Exploiting generic platform technologies for the detection and identification of plant pathogens. Eur J Plant Pathol 121:355–363

Capote N, Pastrana AM, Aguado A, Sánchez-Torres P (2012) Molecular tools for detection of plant pathogenic fungi and fungicide resistance. In: Cumagun CJR (ed) Plant pathology. InTech. ISBN 978-953-51-0489-6, pp 151–202

Cappelli C, Buonaurio R, Pezzotti M, Mazzucato A (1993) A simplification of the embryo test method (ETM) to detect *Ustilago nuda* (Jens.) Rostr. in barley seeds. Phytopathol Mediterr 32:143–144

Cappelli C, Covarelli L (2005) Methods used in seed pathology and their recent improvements. Phytopathol Pol 35:11–18

Carmichael DJ (2012) Developing a sensitive, high-throughput tool for rapid detection of agronomically important seed-borne pathogens of tomato. Master of Science dissertation, University of Witwatersrand, Johannesburg

Chen YY, Conner RL, Gillard CL, McLaren DL, Boland GJ, Balasubramanian PM, Stasolla C, Zhou QX, Hwang SF, Chang KF, Babcock C (2013) A quantitative real-time PCR assay for detection of *Colletotrichum lindemuthianum* in navy bean seeds. Plant Pathol 62:900–907

Cullen DW, Lees AK, Toth IK, Duncan JM (2002) Detection of *Colletotrichum coccodes* from soil and potato tubers by conventional PCR and quantitative real-time PCR. Plant Pathol 51:281–292

De Boer SH, Lopez MM (2012) New grower-friendly methods for plant pathogen monitoring. Annu Rev Phytopathol 50:197–218

De Tempe J, Binnerts J (1979) Introduction to methods of seed health testing. Seed Sci Technol 7:601–636

Dent KC, Stephen JR, Finch-Savage WE (2004) Molecular profiling of microbial communities associated with seeds of *Beta vulgaris* subsp. *vulgaris* (sugar beet). J Microb Methods 56:17–26

Djalali Farahani-Kofoet R, Römer P, Grosch R (2012) Systemic spread of downy mildew in basil plants and detection of the pathogen in seed and plant samples. Mycol Prog 11:961–966

Doyer LC (1938) Manual for the determination of seed-borne diseases. ISTA, Wageningen

Eibel P, Wolf GA, Koch E (2005) Detection of *Tilletia caries*, causal agent of common bunt of wheat by ELISA and PCR. J Phytopathol 153:297–306

EPPO (2007) Diagnostic protocols for regulated pests, PM7/29(2) *Tilletia indica*. EPPO Bull 37:503–520

EPPO (2008) Diagnostic protocols for regulated pests, PM7/85(1) *Plasmopara halstedii*. EPPO Bull 38:343–348

EPPO (2009) Diagnostic protocols for regulated pests, PM7/91(1) *Gibberella circinata*. EPPO Bull 39:298–309

EPPO (2010) Use of EPPO diagnostic protocols, PM7/76 (2). EPPO Bull 40:350–352

EPPO (2013) EPPO A1 and A2 Lists of pests recommended for regulation as quarantine pests. Standard PM 1/2 (22). EPPO, Paris, 16 pp

FAO (2013) ISPM 11 – pest risk analysis for quarantine pests, International Standards for phytosanitary measures. FAO, Rome, 36 pp

Gachon C, Mingam A, Charrier B (2004) Real-time PCR: what relevance to plant studies. J Exp Bot 5:1445–1454

Germini A, Rossi S, Zanetti A, Corradini R, Fogher C, Marchelli R (2005) Development of a peptide nucleic acid array platform for the detection of genetically modified organisms in food. J Agric Food Chem 53:3958–3962

Gil-Serna J, Vázquez C, Sardiñas N, González-Jaén MT, Patiño B (2009) Discrimination of the main Ochratoxin A-producing species in *Aspergillus* section Circumdati by specific PCR assays. Int J Food Microbiol 136:83–87

Guillemette T, Iacomi-Vasilescu B, Simoneau P (2004) Conventional and real-time PCR-based assay for detecting pathogenic *Alternaria brassicae* in cruciferous seed. Plant Dis 88:490–496

Hill NS, Hiatt EE III, De Battista JP, Costa MC, Griffiths CH, Klap J, Thorogood D, Reeves JH (2002) Seed testing for endophytes by microscopic and immunoblot procedures. Seed Sci Technol 30:347–355

Holland PM, Abramson RD, Watson R, Gelfand DH (1991) Detection of specific polymerase chain reaction product by utilizing the 5′–3′ exonuclease activity of *Thermus aquaticus* DNA polymerase. Proc Natl Acad Sci U S A 88:7276–7280

Iacomi-Vasilescu B, Blancard D, Guénard M, Molinero-Demilly V, Laurent E, Simoneau P (2002) Development of a PCR-based diagnostic assay for detecting pathogenic *Alternaria* species in cruciferous seeds. Seed Sci Technol 30:87–95

Ioos R, Fourrier C, Wilson V, Webb K, Schereffer JL, Tourvielle de Labrouhe D (2012) An optimised duplex real-time PCR tool for sensitive detection of the quarantine oomycete *Plasmopara halstedii* in sunflower seeds. Phytopathology 102:908–917

Ioos R, van den Boogert PHHF (2012) EUPHRESCO non-competitive project: ring testing of diagnostic protocols for identification and detection of *Gibberella circinata* in pine seed. Final report. http://www.euphresco.org/downloadFile.cfm?id=730. Accessed 15 Jan 2014

Ioos R, Annesi T, Fourrier C, Saurat C, Chandelier A, Inghelbrecht S, Diogo ELF, Perez-Sierra AM, Barnes AV, Paruma K, Adam M, van Rijswick P, Riccioni L (2013) Test performance study of diagnostic procedures for identification and detection of *Gibberella circinata* in pine seeds in the framework of a EUPHRESCO project. OEPP/EPPO Bull 43(2):267–275

Ioos R, Fourrier C, Iancu G, Gordon TR (2009) Sensitive detection of *Fusarium circinatum* in pine seed by combining an enrichment procedure with a real-time polymerase chain reaction using dual-labeled probe chemistry. Phytopathology 99:582–590

ISTA (2014) International rules for seed testing. Annexe to Chapter 7: Seed Health Testing. http://www.seedtest.org/en/download-ista-seed-health-testing-methods-_content—1–1132–746. html. Accessed 28 Aug 2014

ISTA (2007) ISTA method validation for seed testing V.1. (http://www.seedtest.org/upload/cms/user/ISTAMethodValidationforSeedTesting-V1.01.pdf. Accessed 13 Jan 2014

Josefsen L, Christiansen SK (2002) PCR as a tool for the early detection and diagnosis of common bunt in wheat, caused by *Tilletia tritici*. Mycol Res 106:1287–1292

Kellerer T, Sedlmeier M, Rabenstein F, Killermann B (2006) Development of immunochemical and PCR methods for qualitative detection of *Tilletia* species in organic seeds. Dumalasova, V. (ed.). Special issue. Czech J Genetics Plant Breed 42:72–74

Khanzada AK, Shetty HS, Mathur SB, Cappelli C, Infantino A, Porta-Puglia A (1989) Avoidance of phenol in the embryo count procedure. In: 22nd International Seed Testing Association congress, seed symposium, Edinburgh, 21–30 June 1989. Abstracts of papers. ISTA, Zürich: n. 38

Kochanová M, Zouhar M, Prokinová E, Rysanek P (2004) Detection of *Tilletia controversa* and *Tilletia caries* in wheat by PCR method. Plant Soil Environ 50:75–77

Konstantinova P, Bonants PJM, van Gent-Pelzer M, van der Zouwen P, van den Bulk R (2002) Development of specific primers for detection and identification of *Alternaria* spp. in carrot material by PCR and comparison with blotter and plating assays. Mycol Res 106:23–33

Kubota R, Alvarez AM, Vine BG, Jenkins DM (2007) Development of a loop-mediated isothermal amplification method (LAMP) for detection of the bacterial wilt pathogen *Ralstonia solanacearum* (Abstract). Phytopathology 97:S60

Kulik T, Jestoi M, Okorski A (2011) Development of TaqMan assays for the quantitative detection of *Fusarium avenaceum/Fusarium tricinctum* and *Fusarium poae* esyn1 genotypes from cereal grain. FEMS Microbiol Lett 314:49–56

Lindahl BD, Nilsson RH, Tedersoo L, Abarenkov K, Carlsen T, Kjøller R, Kõljalg U, Pennanen T, Rosendahl S, Stenlid J, Kauserud H (2013) Fungal community analysis by high-throughput sequencing of amplified markers – a user's guide. New Phytol 199:288–299

Ling K, Wechter WP, Walcott RR, Keinath HP (2011) Development of a real-time RT-PCR assay for Squash Mosaic Virus useful for broad spectrum detection of various serotypes and its incorporation into a multiplex seed health assay. J Phytopathol 159:649–656

Majumder D, Rajesh T, Suting EG, Debbarma A (2013) Detection of seed borne pathogens of wheat: recent trends. Aust J Crop Sci 7:500–507

McCartney HA, Foster SJ, Fraaije BA, Ward E (2003) Molecular diagnostics for fungal plant pathogens. Pest Man Sci 59:129–142

McKay GJ, Brown AE, Bjourson AJ, Mercer PC (1999) Molecular characterisation of *Alternaria linicola* and its detection in linseed. Eur J Plant Pathol 105:157–166

McNeil M, Roberts AMI, Cockerell V, Mulholland V (2004) Real-time PCR assay for quantification of *Tilletia caries* contamination of UK wheat seed. Plant Pathol 53:741–750

Mumford R, Boonham N, Tomlinson J, Barker I (2006) Advances in molecular phytodiagnostics – new solutions for old problems. Eur J Plant Pathol 116:1–19

Munkvold GP (2009) Seed pathology progress in academia and industry. Annu Rev Phytopathol 47:285–311

Neergaard P (1979) Seed pathology, vols I–II, 2nd edn. MacMillan Press, London/Basingstoke

Niessen L, Vogel RF (2010) Detection of *Fusarium graminearum* DNA using a loop-mediated isothermal amplification (LAMP) assay. Int J Food Microbiol 140:183–191

Noble M, de Tempe J, Neergaard P (1958) An annotated list of seed-borne diseases. Commonwealth Mycological Institute, Kew

Noble M, Richardson MJ (1968) An annotated list of seed-borne diseases. 2nd ed. Proc Int Seed Test Assoc 33:1–91

Notomi T, Okayama H, Masubuchi H, Yonekawa T, Watanabe K, Amino N, Hase T (2000) Loop-mediated isothermal amplification of DNA. Nucl Acid Res 28(12):e63

OECD (2012) A synthesis of international regulatory aspects that affect seed trade. OECD Seed Schemes, http://www.oecd.org/tad/code/internationalregulatoryaspectsseedtrade.pdf. Accessed 28 Jan 2014

Parida MM, Santhosh SR, Dash PK, Lakshmana Rao PV, Morita K (2008) Rapid and real-time assays for detection and quantification of Chikungunya virus. Future Virol 3:179–192

Pasquali M, Piatti P, Gullino ML, Garibaldi A (2006) Development of a real-time polymerase chain reaction for the detection of *Fusarium oxysporum* f. sp. *basilici* from basil seed and roots. J Phytopathol 154:632–636

Pellegrino C, Gilardi G, Gullino ML, Garibaldi A (2010) Detection of *Phoma valerianellae* in lamb's lettuce seeds. Phytoparasitica 38:159–165

Peters J, Sledz W, Bergervoet JHW, van der Wolf JM (2007) An enrichment microsphere immunoassay for the detection of *Pectobacterium atrosepticum* and *Dickeya dianthicola* in potato tuber extracts. Eur J Plant Pathol 17:97–107

Quarta A, Mita G, Haidukowski M, Logrieco A, Mulè G, Visconti A (2006) Multiplex PCR assay for the identification of nivalenol, 3- and 15-acetyl-deoxynivalenol chemotypes in *Fusarium*. FEMS Microbiol Lett 259:7–13

Richardson MJ (1979) An annotated list of seed-borne diseases, 3rd edn. CMI, Kew and ISTA, Zurich

Roberts AMI, Theobald CM, McNeil M (2007) Calibration of quantitative PCR assays. J Agric Biol Environ Stat 12:364–378

Stobbe AY, Daniels J, Espindola AS, Verma R, Melcher U, Ochoa-Corona F, Garzon C, Fletcher J, Schneider W (2013) E-probe diagnostic nucleic acid analysis (EDNA): a theoretical approach for handling of next generation sequencing data for diagnostics. J Microb Methods 94:356–366

Szemes M, Bonants P, deWeerdt M, Baner J, Landegren U, Schoen CD (2005) Diagnostic application of padlock probes: multiplex detection of plant pathogens using universal microarrays. Nucl Acid Res 33:e70

Tsui CKM, Woodhall J, Chen W, LéVesque CA, Lau A, Schoen CD, Baschien C, Najafzadeh MJ, De Hoog GS (2011) Molecular techniques for pathogen identification and fungus detection in the environment. IMA Fungus 2:177–189

Tsui CKM, Wang B, Schoen CD, Hamelin RC (2013) Rapid identification and detection of pathogenic fungi by padlock probes. In: Gupta VK et al (eds) Laboratory protocols in fungal biology: current methods in fungal biology; fungal biology. Springer, New York/Heidelberg/Dordrecht/London

Tucker T, Marra M, Friedman JM (2009) Massively parallel sequencing: the next big thing in genetic medicine. Am J Hum Genet 85:142–154

Udayashankar AC, Chandra Nayaka S, Archana B, Anjana G, Niranjana SR, Mortensen CN, Lund OS, Prakash HS (2012) Specific PCR-based detection of *Alternaria helianthi*: the cause of blight and leaf spot in sunflower. Arch Microbiol 194:923–932

van Doorn R, Szemes M, Bonants P, Kowalchuk GA, Salles JF, Ortenberg E, Schoen CD (2007) Quantitative multiplex detection of plant pathogens using a novel ligation probe-based system coupled with universal, high-throughput real-time PCR on OpenArrays®. BMC Genetics 8:276

Vincelli P, Tisserat N (2008) Nucleic acid-based pathogen detection in applied plant pathology. Plant Dis 92:660–669

Waalwijk C, Kastelein P, de Vries I, Kerènyi Z, van der Lee T, Hasselink T, Köhl J, Kema GHJ (2003) Mjr changes in *Fusarium* spp. in wheat in the Netherlands. Eur J Plant Pathol 109:743–754

Waalwijk C, van der Heide R, de Vries I, van der Lee T, Schoen C, Costrel-Decorainville G, Haeuser-Hahn I, Kastelein P, Köhl J, Lonnet P, Demarquet T, Kema GHJ (2004) Quantitative detection of *Fusarium* species in wheat using TaqMan. Eur J Plant Pathol 110:481–494

Walcott R, Gitaitis RD, Langston DB (2004) Detection of *Botrytis aclada* in onion seed using magnetic capture hybridization and the polymerase chain reaction. Seed Sci Technol 32:425–438

Walcott RR (2003) Detection of seed-borne pathogens. Hortic Technol 13:40–47

Walcott RR, Ha Y, Johnson K (2008) Simultaneous detection of multiple pathogens in seeds using magnetic capture hybridization and real-time PCR. J Plant Pathol 90:S2.209

White JR, Maddox C, White O, Angiuoli SV, Fricke WF (2013) CloVR-ITS: Automated internal transcribed spacer amplicon sequence analysis pipeline for the characterization of fungal microbiota. Microbiome. doi:10.1186/2049-2618

Woese CR (2004) A new biology for a new century. Microbiol Mol Biol Rev 68:173–186

Zhang N, Geiser DM, Smart CD (2007) Macroarray detection of solanaceous plant pathogens in the *Fusarium solani* species complex. Plant Dis 91:1612–1620

Zouhar M, Mazáková J, Prokinová E, Vánová M, Rysánek P (2010) Quantification of *Tilletia caries* and *Tilletia controversa* mycelium in wheat apical meristem by real-time PCR. Plant Prot Sci 46:107–115

Part III
Seed Treatments

Chapter 7
Benefits of Chemical Seed Treatments on Crop Yield and Quality

Gary P. Munkvold, Clifford Watrin, Monika Scheller, Ronald Zeun, and Gilberto Olaya

Abstract Agriculturists have been treating seeds to protect them from pathogens and pests for centuries, even before the nature of plant diseases was understood. Today, the use of seed treatments is a highly sophisticated strategy that has evolved into a very valuable, effective, and environmentally friendly component of agricultural production practices. Chemical seed treatments can be used to achieve a variety of benefits, including: improved emergence, through protection from seedborne pathogens and soilborne pathogens and insects; prevention of seed transmission of seedborne pathogens; protection of above-ground plant parts from infection by airborne pathogens or feeding by insect pests and disease vectors; improved vigor and uniformity of crop growth; deterrence of deterioration or insect feeding in storage; fulfillment of phytosanitary requirements and prevention of pathogen spread. These benefits all contribute to maximizing crop yield and quality while minimizing negative impacts through efficient use of crop protection chemicals. Seed treatment allows for highly targeted application of low, uniform doses of product, which is effective while reducing the risk of selection pressure for pathogen or pest resistance. Seed treatments are commercially available with contact, locally systemic, or fully systemic activity. Common active ingredients can be used for protection against Oomycetes, fungi, insects, and nematodes. There are numerous examples of improvements in stand establishment and yield as a result of seed treatment use in a wide range of crops. Combinations of active ingredients are becoming more common as products improve for efficacy against specific pathogen groups. In maize, seed treatment is nearly universal and standard practices may include a combination of four fungicides, an insecticide, and a

G.P. Munkvold (✉)
Iowa State University, Ames, IA, USA
e-mail: munkvold@iastate.edu

C. Watrin
Syngenta Crop Protection, Stanton, MN, USA

M. Scheller
Syngenta Crop Protection, Basel, Switzerland

R. Zeun
Syngenta Crop Protection, Stein, Switzerland

G. Olaya
Syngenta Crop Protection, Vero Beach, FL, USA

M.L. Gullino, G. Munkvold (eds.), *Global Perspectives on the Health of Seeds and Plant Propagation Material*, Plant Pathology in the 21st Century 6,
DOI 10.1007/978-94-017-9389-6_7

nematicide. This provides a high level of protection across a wide pathogen spectrum as well as prevention of feeding damage to the seed and seedling. Seed treatments are playing an increasing role in the productivity of agriculture, as well as its sustainability and efficiency. Seed application of crop protection compounds provides unique benefits that make it a preferable approach compared to other tactics. It is a reliable technology that guarantees a uniform crop establishment in a variety of environments, soils and cultural practices; benefits provided by seed treatments cannot be duplicated because most of the target diseases and pests cannot be controlled after planting.

Keywords Fungicide • Insecticide • Nematicide • Seed transmission • Seedling disease • Damping-off

1 Introduction

Seed treatment is one of the oldest known tactics for management of plant diseases (Russell 2005). Seed treatments, in some form, have been used for centuries, and used widely on a commercial basis for decades. But during the last 10 years, seed treatment use has rapidly accelerated and evolved, and seed treatment is now an integral component of management strategies for soil and seedborne pathogens, nematodes, and insect pests. Many factors are driving this growth. Crop productivity must be maximized in order to meet global demand for food and fuel, while the increased use of foliar and soil applications of crop protection chemicals is undesirable due to environmental concerns. Emergence or re-emergence of diseases and pests in new locales is adding to the need for additional crop protection tools. For economic and logistical reasons, crop production practices have moved toward tactics that increase the risks for seed decay, seedling disease, and early-season insect attack. Innovations in seed treatment technology allow for precise application of active ingredients, offering high levels of efficacy while reducing environmental exposure. These factors and others have resulted in rapid growth of the global seed treatment market, from approximately USD $1 billion in 2002 to more than $3 billion in 2012, with forecasted growth to more than $4 billion by 2017.

This rapid growth in the popularity of seed treatments also is part of a broader trend in agriculture that emphasizes the value of seed, and the potential of seed as a delivery mechanism for crop management inputs. The central role of seeds in agriculture has always been recognized, but the importance of this role has been greatly heightened during the past century and especially the last decade. Several developments have catalyzed the elevation of seed as the most valuable agricultural input. In several of the world's major crops, the development and implementation of hybrids has resulted in a major emphasis on seed production practices. Improved breeding methods, including the use of marker-assisted selection and other so-called "molecular breeding" techniques have enhanced the importance of producing and distributing improved cultivars that deliver high levels of yield, quality, and stress tolerance. Intellectual property protection such as the U.S. Plant Variety Protection Act has contributed to recognition of the value of improved cultivars distributed as seed. The advent of biotechnology has further promoted the value of seed due to the incorporation of valuable pest management traits into the seed; this dimension will only accelerate as other traits with high value to the consumer are added to the repertoire of genetically modified crop plants. The role of seed treatments will continue to expand as crop producers seek to protect their growing investment in high-value seed, and expect more and more input and output traits to be delivered with the seed.

Seed treatment can be physical, chemical, or biological. Chemical and biological seed treatments may have a wide variety of objectives, but in this chapter we will focus on the application of bioactive chemicals to the seed prior to sowing, with the primary purpose of deterring damage by fungal and Oomycete plant pathogens, insect pests, and nematodes. Seed treatment can be used to achieve the following benefits:

- Improve emergence and promote stand establishment by neutralizing seedborne pathogens and protecting against soilborne pathogens and insects
- Prevent seed transmission of seedborne pathogens
- Protect above-ground plant parts from infection by airborne pathogens or feeding by insect pests and disease vectors
- Promote uniform growth of the crop
- Maximize crop yields
- Deter deterioration or insect feeding in storage
- Meet phytosanitary requirements and prevent spread of pathogens

Seed treatment offers several advantages as a tactic for disease and pest management:

Product right on the target – seed-applied active ingredients are in the right place at the right time, protecting the seed and seedling roots during the first critical stages of crop development. Spray or granular applications to soil result in a large proportion of product failing to come into contact with target organisms.

Uniform plant-per-plant loading – modern seed treatment application technology allows for precise dosing of active ingredients down to 0.001 mg active ingredient or less per seed. Each plant receives the desired dose of active ingredient, regardless of plant spacing or planting density, and without the variability inherent in field applications.

Continuous delivery from depot in soil – diffusion of the active ingredient into the spermosphere and rhizosphere can continue for weeks after planting, and this can be modulated through the use of polymer coatings.

Reduction of product load per ha – In order to achieve the same results, spray and granular applications require much larger quantities of active ingredient and other formulation components, largely because of lower precision in placement of the product. For example, to control soilborne insects in maize, a typical soil application of a sprayed product would require more than 1 kg/ha of active ingredient, and granular applications typically require 250–600 g/ha, whereas seed-applied insecticides are used at rates equivalent to 40–100 g/ha. The field area exposed to the active ingredient also is much smaller with seed-applied products; a spray treatment covers the entire field area, an in-furrow granular treatment would expose approximately 500 m^2/ha to the products, but a seed-applied product would contact only about 58 m^2/ha.

Convenient Application – application to the seed at the seed production or retail facility allows for the centralization of crop protection chemical application, avoiding the equipment costs and inconvenience, and drastically reducing risk of exposure at the farmer level. Application of crop protection products to seeds is conducted in a well-controlled, contained, safe manner.

Reduced risk for pathogen resistance – seed treatments are less likely to cause selection pressure that leads to fungicide resistance in soilborne pathogen populations. Because most of the population is not exposed to the active ingredient, selection pressure is greatly reduced with seed treatments. However, this benefit can be negated if the same active ingredient (or others in the same FRAC group) are used as foliar or soil applications in the same field.

2 History of Seed Treatment Use

Seed treatment has evolved significantly from its humble beginnings, when cereal seeds were soaked in brine to prevent the occurrence of smut diseases. Once it became clear that chemical treatments could improve the health of seeds and prevent diseases, a succession of active ingredients was implemented as seed treatments. Initially the products used were very broad-spectrum chemicals with undesirable toxicological profiles, but increasingly sophisticated products with specific activities, lower rates and reduced mammalian toxicities have been developed and applied to seeds (Table 7.1). The trend toward lower-rate chemistries has been very clear, beginning in the 1990s (Fig. 7.1).

Use of seed treatments also has grown across a wide range of crops. Initially limited to cereal crops and maize, seed treatments are now applied to a high percentage of soybeans, other diverse field crops, specialty crops, and vegetables (Fig. 7.2).

Table 7.1 Evolution of seed treatment chemicals introduced from the 1600s to early 2000s

Time period	Seed treatments introduced
1600s–1700s	Brine, arsenic, copper sulfate
1920s	Organic compound (organo-mercury)
1930s	Thiram, Terraclor (PCNB)
1950s	Captan
1960s	Systemic fungicides for seedborne diseases (e.g., carboxin, thiabendazole)
1970s and 1980s	Systemic fungicides for airborne diseases (e.g., triadimenol, ethirimol) and seedborne diseases (imazalil)
1990s	Broad spectrum, low-rate compounds – fludioxonil, tebuconazole, triticonazole, difenoconazole
2000s	QoI fungicides – azoxystrobin, trifloxystrobin, pyraclostrobin, others
2010s	SDHI fungicides – sedaxane, fluxapyroxad, fluopyram, penflufen, others

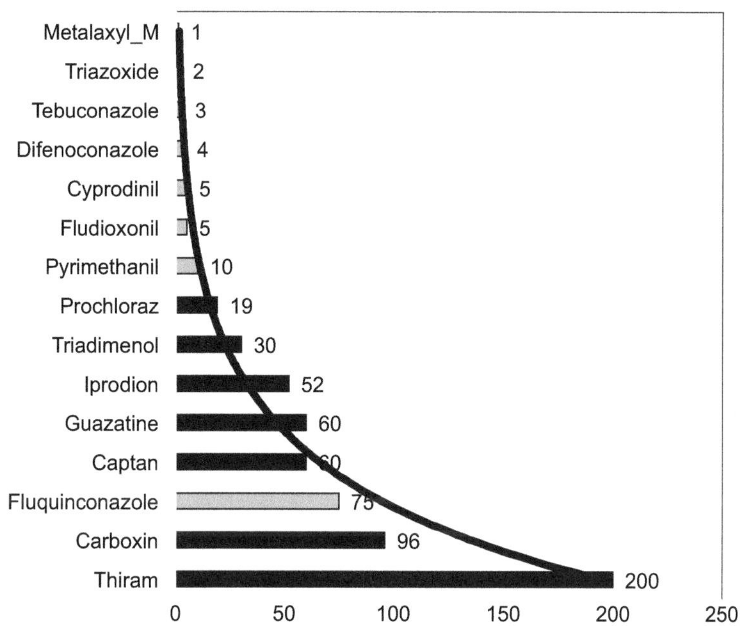

Fig. 7.1 Typical application rates (g active ingredient per 100 kg seed) for seed treatment fungicides introduced prior to the 1990s (*black* bars) vs. those introduced during the 1990s or later (*gray* bars)

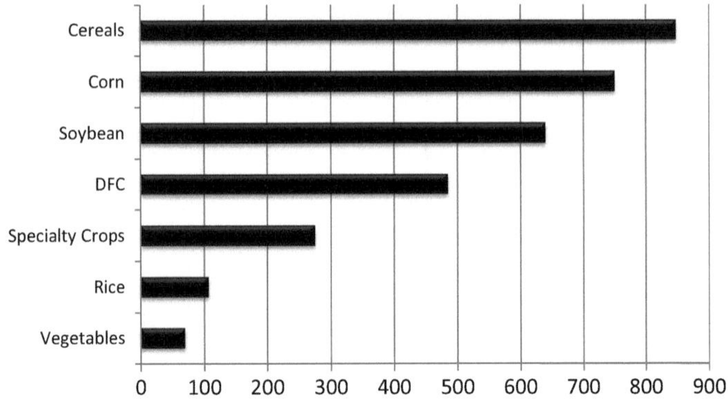

Fig. 7.2 Global seed treatment sales in million USD for different crop markets in 2012 (Data from Syngenta Seed Care, 2013). *DFC* diverse field crops (sugar beets, sunflower, oil seed rape); Specialty crops: Potato, cotton, peanut, fruit, coffee and other crops

3 Activity of Chemical Seed Treatments

The scope of the benefits that can be achieved with a seed treatment depends on the type of activity that the active ingredient possesses. Traditionally, this has been defined in terms of how the target pathogen or pest is exposed to the bioactive chemical.

(a) Contact activity – the bioactive chemical is not taken up by plant tissues; the target organism is exposed to the chemical only by direct contact with the chemical on the seed surface or in the soil as the chemical diffuses from the seed coat into the spermosphere and rhizosphere. Common examples are captan (fungicide), fludioxonil (fungicide), and lindane (insecticide).

(b) Locally systemic – the bioactive chemical is taken up by the plant tissues which come in contact with it, but the chemical is not translocated within the plant. Common examples are some of the QoI fungicides (e.g., trifloxystrobin).

(c) Systemic – the bioactive chemical is taken up by plant tissues and translocated within the plant. Common examples are mefenoxam (for Oomycetes), difenoconazole (fungicide), and thiamethoxam (insecticide).

In addition to these categories of activity, seed treatment chemicals also may have activity that protects plants from pathogens indirectly, by altering the physiology of the plant. The most well-known example of this type of activity is systemic acquired resistance (SAR). SAR represents the triggering of the plants' chemical defense mechanisms in response to a stimulus – in this case, a chemical. An example of an SAR inducer is acibenzolar-S-methyl (Actigard®, Bion®) from Syngenta. Acibenzolar-S-methyl induced a broad resistance spectrum on several mococotyledonous and dicotyledonous plants. Acibenzolar-S-methyl is one of the most widely investigated molecules as a positive marker of SAR in various species of plants, using diversified research approaches in many countries (Toquin et al. 2012). Harpin, a bacterial-derived protein, is another example of an SAR inducer. Harpin is used as a seed treatment (N-hibit™, Plant Health Care, Inc.).

4 Seed Treatment Benefits for Management of Diseases, Insects, and Nematodes

Seedborne and soilborne diseases – Modern commercial seed treatments typically consist of mixtures of fungicides with complementary spectra of activity to cover a wide range of fungi. This is because there are multiple threats to the seed and seedling – Oomycetes, Ascomycetes, Basidiomycetes, etc. For example, a standard fungicide seed treatment package for maize such as Maxim®Quattro (Syngenta Seed Care, Basel, Switzerland) includes four fungicides: mefenoxam (systemic Oomycete product), fludioxonil (broad-spectrum contact fungicide), azoxystrobin (broad-spectrum systemic fungicide), and thiabendazole (broad-spectrum systemic fungicide, particularly for *Fusarium* spp.). This combination improves germination due to control of seedborne fungi such as *Fusarium* spp., and protects the seed from soilborne fungi and Oomycetes for several weeks after planting. In trials conducted in various U.S. locations in 2010, this combination increased plants/ha by more than 15 % and improved yield by more than 7 % compared to untreated seed (Syngenta 3rd-party data, 2010) (Fig. 7.3). Other commercial combinations for maize include metalaxyl (systemic Oomycete product) combined with trifloxystrobin (broad-spectrum locally systemic fungicide) (Bayer Crop Science) or metalaxyl and trifloxystrobin combined with ipconazole (broad-spectrum systemic fungicide) (Acceleron®, Monsanto Co., St. Louis, MO).

While seed treatment has been universal for years on some crops, such as maize, in other crops, such as soybean, adoption has been more recent. Before 2000, less than 10 % of the U.S. soybean crop received a seed treatment, but adoption increased to over 75 % by 2013. Various factors have driven this increase; it is sustained primarily because of positive results. For example, Esker and Conley (2012) found that soybean seed treatment with mefenoxam + fludioxonil or mefenoxam + fludioxonil + thiamethoxam consistently provided a positive yield response in field trials at multiple locations and years in Wisconsin, USA. They calculated probabilities for a positive return on investment and found that seed treatment provided positive returns in the majority of cases, depending on soybean cultivar, grain price, and yield environment.

Oomycetes are a nearly universal target for seed treatments across all crops, because of the diversity of genera and species of Oomycetes, and their adaptation to cause infection of seeds and seedlings during early-season conditions. Metalaxyl and mefenoxam have been the mainstays for seed treatment products against *Pythium* spp. and *Phytophthora* spp., and they are effective, but research is very active for discovery of new active ingredients with Oomycete activity. In some

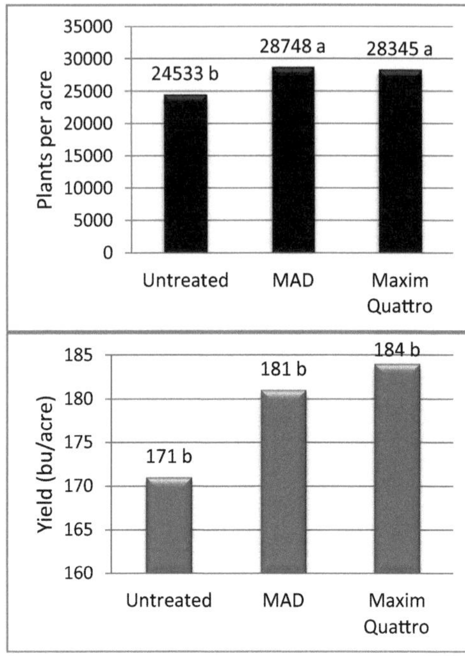

Fig. 7.3 Effects of fungicide seed treatments on plant stands and yield of maize, 2010. MAD = fludioxonil + mefenoxam + azoxystrobin. Maxim®Quattro = fludioxonil + mefenoxam + azoxystrobin + thiabendazole (Syngenta 3rd-party data, 2010)

crops, *Aphanomyces* spp. are important seedling pathogens, and they are not well-controlled by metalaxyl or mefenoxam. Hymexazol is a unique seed treatment product with efficacy against *Aphanomyces* and *Pythium*. It has been successfully used against *A. cochlioides* in sugar beets, and can drastically improve sugar beet stands under high disease pressure, maintaining 80–90 % seedling survival when untreated seeds have less than 10 % survival (Harveson et al. 2007).

The recent introduction of commercial seed treatment products including SDHI (succinate-dehydrogenase inhibitor) fungicides offers the potential to improve control of Basidiomycete fungi, compared to currently available seed treatments. This is valuable because of the importance of *Rhizoctonia solani* and other *Rhizoctonia* species as soilborne seedling pathogens in many crops, and because of the importance of seedborne smut diseases in cereals. In soybean field trials in various U.S. locations, the use of sedaxane seed treatment (combined with mefenoxam and fludioxonil) increased yields by an average of 14 % over the untreated control and 7.8 % over the yield from seeds treated only with mefenoxam and fludioxonil. Yield increases were even greater in trials inoculated with *R. solani* (27 % higher than untreated and 8.7 % higher than mefenoxam/fludioxonil) (Fig. 7.4). Similar yield increases also have been reported for control of *R. solani* with sedaxane in maize. In wheat, sedaxane is combined with difenoconazole and mefenoxam, and demonstrated 15.5 % yield increase over untreated seed in natural-infested field trials (Syngenta 3rd-party data, 2013). Penflufen, another SDHI fungicide, is used as a

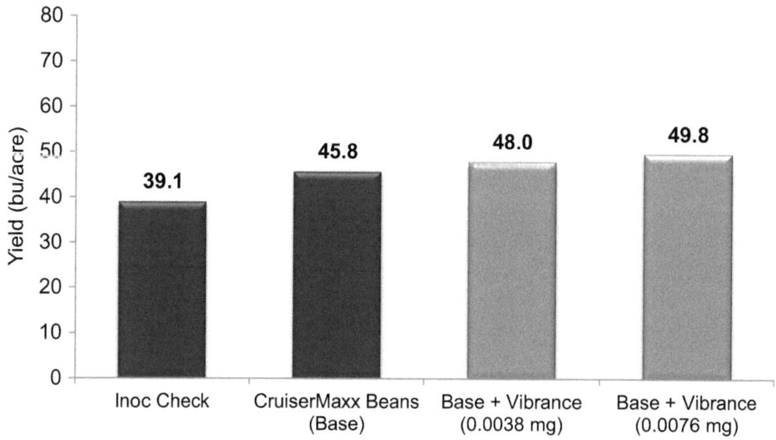

Fig. 7.4 Soybean yield increases with sedaxane (Vibrance®) seed treatment in inoculated field trials, 2011–2012 (Data from Syngenta Seed Care, 2013)

seed treatment in soybean and wheat, and also has shown significant yield effects due to *R. solani* control in both crops. Smut diseases affect maize and cereal crops such as barley. In maize, seed treatment with sedaxane combined with fludioxonil, mefenoxam, and azoxystrobin reduced head smut (*Sphacelotheca reiliana*) symptoms from 31.1 % down to 4.5 % (Syngenta data, 2013). In barley, loose smut (*Ustilago nuda*) was 95–100 % controlled by seed treatment with penflufen (Bayer Crop Science, 2013).

Seedling pathogens such as *Fusarium* spp. also can persist in maize plant tissues and cause symptoms later in the season, including crown rot and stalk rot. In some cases, seed treatment fungicides have demonstrated reductions in these symptoms. Rodriguez-Brljevich et al. (2009) showed that seed treatment with a combination of the fungicides fludioxonil, mefenoxam, azoxystrobin, and the insecticide thiamethoxam resulted in significant reductions in root rot symptoms, crown rot symptoms, stalk rot, and the incidence of *Fusarium* infection of maize roots.

Airborne fungal pathogens – The introduction of several new systemic fungicides in the 1970s opened a new dimension in seed treatment benefits through translocation of fungicide active ingredients to above-ground plant parts. This has allowed for protection against diseases that affect leaves of seedlings, such as downy mildew, leaf rust and stem rust, Stagonospora leaf blotch, and soybean rust. The benefit of seed treatment for these diseases is primarily to delay the onset of epidemics by suppressing disease during the early seedling growth stages. This period is critical for establishment of a vigorous stand, and delaying epidemic onset reduces the overall impact of these foliar diseases. Sundin et al. (1999) found that sporulation of *Septoria tritici*, *Stagonospora nodorum*, and *Puccinia recondita* on wheat seedling leaves was suppressed for several weeks by seed treatment with difenoconazole or triadimenol. Similarly, stripe rust (*Puccinia striiformis*) symptoms were suppressed for 2–3 weeks with difenoconazole or triadimenol seed treatment (Chen 2005). *Stagonospora nodorum* symptoms were suppressed through growth stage 4 on seedlings from difenoconazole or triadimenol-treated seeds in Arkansas (Milus and Chalkley 1997). In Brazil, fluquinconazole seed treatment similarly delayed the development of Asian soybean rust symptoms (Goulart et al 2011).

Insect pests - The most dramatic change in seed treatment use during the past 15 years has been the rapid adoption of seed-applied insecticides in several crops. The popularity of seed-applied insecticides revolves around the neonicotinoid active ingredients. Although some insecticides were approved and marketed as seed treatments prior to the introduction of the neonicotinoid insecticides, their use was very limited. Imidacloprid was introduced as a seed treatment for maize in 1995, followed by thiamethoxam (1997 in New Zealand; 2001 in the U.S.) and clothianidin (2003). Since 2000, the use of these products as seed treatments has increased dramatically, and currently either thiamethoxam or clothianidin is used as a standard seed treatment for more than 90 % of the maize seed planted in the U.S. This trend has occurred in other crops as well. For example, in sugar beet in the United Kingdom, use of seed-applied insecticides went from zero in 1993 to about 75 % of the area sown to sugar beets in 2002 (Dewar et al. 2003). This corresponded with a dramatic 95 % drop in overall insecticide use on sugar beets in the UK, as seed treatment replaced soil-applied insecticides. The same seed-applied insecticides also are now used on a majority of canola seed planted in North America, and on increasing percentages of soybean and cotton seed. In maize, these products were initially intended to contribute to control of corn rootworms (*Diabrotica* spp.) along with other pests such as wireworm (*Melanotus* spp.), seed corn maggot (*Hylemya platura*), and black cutworm (*Agrotis ypsilon*). Corn rootworm management strategies are now dominated by maize hybrids with transgenic resistance to corn rootworms; this has led to substantial reductions in insecticide use on maize and the substitution of the transgene/seed-applied insecticide combination in place of soil-applied insecticides. Seed-applied clothianidin or imidacloprid have shown excellent control of seed corn maggot, black cutworm, white grub (*Popillia japonica*), grape colaspis (*Colaspis brunnen*), and wireworm, providing equivalent or better results than in-furrow insecticides (Andersch and Schwarz 2003).

Although seed-applied insecticides are used primarily for control of soilborne insects, their systemic properties have led to significant contributions to management of insect-vectored diseases. Because several aerial insects are vectors of plant pathogens, in some cases seed-applied insecticide use has contributed to reductions in disease transmission. In maize, Stewart's wilt, caused by the bacterium *Pantoea stewartii*, is an important quarantined pathogen that can be seed-transmitted. Therefore it is important to minimize the occurrence of the disease in maize seed production fields. Stewart's wilt also causes economic losses, especially in sweet corn, by prematurely killing plants and blighting leaves. Seed parent plants that are infected early in their development are more likely to produce infected seeds (Block et al. 1999), and sweet corn plants infected early in their development are more likely to die or suffer yield loss through leaf blighting (Pataky et al. 1995;

Suparyono and Pataky 1989), so seedling protection against the insect vector (corn flea beetle) has been an important management component of maize seed production and sweet corn production. Seed-applied neonicotinoid insecticides have consistently been shown to effectively prevent feeding by the corn flea beetle and significantly reduce transmission of *P. stewartii* (Andersch and Schwarz 2003; Kuhar et al. 2002; Munkvold et al. 1996; Pataky et al. 2000; Pataky et al. 2005). For example, in field experiments conducted from 2000 through 2003, Pataky et al. (2005) showed average reductions in Stewart's wilt incidence in sweet corn were 75.5 % for clothianidin (0.19–0.25 mg a.i./seed), 69.6 % for imidacloprid (0.34 mg a.i./seed), and 69.3 % for thiamethoxam (0.25–0.27 mg a.i./seed). In cantaloupe, seed-applied imidacloprid reduced the severity of bacterial wilt, caused by *Erwinia tracheiphila*, through control of its vector, the striped cucumber beetle (Fleischer et al. 1998). Seed applied insecticides also can reduce aphid transmission of viruses in oats, sorghum, sugarbeet, and wheat (Maienfisch et al. 1999). Gourmet et al. (1996) showed reductions in the spread of *Barley yellow dwarf virus* in oats and wheat when seed was treated with imidacloprid; similarly, Harvey et al. (1996) demonstrated reduced incidence of *Sugarcane mosaic virus* strain MDMV-B in sorghum with imidacloprid seed treatment. In sugarbeet, both *Beet mild yellowing virus* and *Beet yellows virus* incidence were reduced by seed treatment with imidacloprid or clothianidin (Dewar et al. 2003). Control of wheat curl mite resulted in reduced incidence of *Wheat streak mosaic virus* in wheat (Harvey et al. 1998). In soybeans, seed-applied insecticides can reduce spread of *Bean pod mottle virus* (BPMV) by overwintering bean leaf beetles (Daniels et al. 2001). Recommendations for integrated control of bean leaf beetle and BPMV call for the use of a seed-applied insecticide or seedling-stage foliar application (Rice et al. 2007). Seed-applied thiamethoxam can reduce soybean aphid damage (McCornack and Ragsdale 2006), which would seem to have potential to reduce spread of *Soybean mosaic virus* by aphid vectors.

Through the combination of corn rootworm control, control of other soilborne insects, and the contribution to managing insect-vectored diseases, neonicotinoid insecticide seed treatments have a dramatic economic impact on agriculture. In a study evaluating the impacts of neonicotinoid insecticides in Europe, Noleppa and Hahn (2013) estimated that the annual value of neonicotinoid insecticides across wheat, barley, maize, sunflower, oilseed rape, and sugar beet in 10 EU countries was 2.8 billion EUR in agriculture, with an economy-wide value of 3.8–4.5 billion EUR. These authors concluded that, "Neonicotinoid seed treatment is a key and currently often irreplaceable technology available to farmers today that helps secure the competitiveness of European agriculture – with all the discussed socio-economic benefits and global environmental benefits – as well as achieve a level of productivity that supports the stability of agricultural markets, while also supporting the food security for a growing world population. The authors strongly recommend that these facts are considered in any regulatory decision-making process that addresses this technology."

Nematicidal seed treatments – Another recent development in seed treatment use has been the implementation of nematicidal seed treatments. Plant-parasitic nematodes feed on every crop to some extent, and their impact is highly variable. Nematode feeding during the seedling stage can delay crop development and reduce yield potential, and also can enhance the severity of other seedling diseases. Chemical control of nematodes has traditionally been accomplished through soil fumigation or soil application of nematicides. However, these practices are not often cost-effective. Seed-applied nematicides are now available for several crops including maize, soybean, and cotton; they are based on active ingredients abamectin (marketed in the U.S. as Avicta®, Syngenta Seed Care) or thiodicarb (marketed in the U.S. as Aeris®, Bayer Crop Science). Both are broad-spectrum nematicides with activity against all plant-parasitic nematodes. Both are marketed together with an insecticide product and are usually combined with a fungicide or fungicide combination. Target nematode species include root-lesion, root-knot, reniform, and others. When nematodes are at damaging levels, abamectin has been shown to increase maize stands and yields by more than 5 %, with a multi-year average yield response of 6 bu/acre or 377 kg/ha (data from Syngenta Seed Care, 2013). This represents a return on investment of at least 3:1. In laboraotry research, interactions between root-lesion nematodes (*Pratylenchus penetrans*) and fungal pathogens of seedlings led to increased seedling disease that was effectively managed with a combination of abamectin and seed treatment fungicides (da Silva 2011). Aeris® and Avicta® seed treatments have been shown to increase cotton yields by 6–12 % or more through reductions in feeding by reniform and other nematodes, with results equivalent to or superior to Temik applications to soil (data from Bayer Crop Science, 2007). Biological seed treatment products are now available for nematode control, but this is beyond the scope of this chapter.

Effects on plant vigor – following the adoption of neonicotinoid seed treatments, observations of enhanced crop growth and yield were often made when insect populations were not believed to be damaging. This led to an increased focus on the effects of seed treatments on plant vigor, which has proven to be a tangible benefit to seed treatment use in some cases.

Strobilurin fungicides have been shown to induce physiological changes in plants (Bartlett et al. 2002), including suppression of ethylene biosynthesis, increased levels of abscisic acid (Grossmann et al. 1999), and enhanced antioxidative potential (Wu and Von Tiedemann 2001), resulting in delayed senescence or prolonged leaf greenness (Grossmann and Retzlaff 1997), increased tolerance to environmental stresses (Beck et al. 2002), improved CO_2 and Nitrogen assimilation (Glaa and Kaiser 1999), and increased water use efficiency due to reduced transpiration. At least one report also cites induction of plant-defense responses in treated seed, resulting in better resistance to virus diseases (Herms et al. 2002). Effects on plant physiology have been documented for triazoles and other fungicides previously (Tripathi et al. 1980), but the value of these physiological responses has only recently been considered regarding seed treatment use.

Neonicotinoid seed treatment insecticides also can have beneficial physiological effects on plants. Positive effects on plant establishment, growth and yield have been associated with the neonicotinoid insecticides, particularly thiamethoxam (Mutton et al. 2007; Prasanna et al. 2004). This effect has been documented in numerous crops, including canola, rice, potatoes, maize, soybean, peas, sugar beet, cotton, sugarcane, and sunflowers (Aramaki et al. 2013). Although there is abundant evidence for positive physiological effects of some fungicide and insecticide seed treatments on plants, the magnitude of physiological effects is variable as are their economic impacts on crop productivity (Bertelsen et al. 2001; Nason et al. 2007; Palumbo and Sanchez 1995).

Conclusion
Seed treatments are playing an increasing role in the productivity of agriculture, as well as its sustainability and efficiency. Seed application of crop protection compounds provides unique benefits that make it a preferable approach compared to other tactics. It is a reliable technology that guarantees a uniform crop establishment in a variety of environments, soils and cultural practices; benefits provided by seed treatments cannot be duplicated because most of the target diseases and pests cannot be controlled after planting.

References

Andersch W, Schwarz M (2003) Clothianidin seed treatment (Poncho®) – the new technology for control of corn rootworms and secondary pests in US-corn production. Pflanzenschutz-Nachrichten Bayer 56:147–172

Aramaki PH, da Silva AJ, Castro PRC (2013) Crop enhancement: releasing plant potential. Syngenta Crop Protection, Santo Amaro-SP, p 192 p

Bartlett DW, Clough JM, Godwin JR, Hall AA, Hamer M, Parr-Dobrzanski B (2002) The strobilurin fungicides. Pest Manag Sci 58:649–662

Beck C, Oerke EC, Dehne HW (2002) Impact of strobilurins on physiology and yield formation of wheat. Meded Rijksuniv Gent Fak Landbouwkd Toegep Biol Wet 67:181–187

Bertelsen JR, de Neergaard E, Smedegaard-Petersen V (2001) Fungicidal effects of azoxystrobin and epoxiconazole on phyllosphere fungi, senescence and yield of winter wheat. Plant Pathol 50:190–205

Block CC, McGee DC, Hill JH (1999) Relationship between late season Stewart's bacterial wilt and seed infection in maize. Plant Dis 83:527–530

Chen XM (2005) Epidemiology and control of stripe rust (*Puccinia striiformis* f. sp. *tritici*) on wheat. Can J Plant Pathol 27:314–337

Da Silva MP (2011) Interactions between root-lesion nematodes and corn pathogens. MS thesis. Iowa State University, Ames, 109 pp

Daniels JL, Munkvold GP, McGee DC (2001) Comparison of infected soybean seed and bean leaf beetles as inoculum sources for *Bean pod mottle virus* (Abstract). Phytopathology 91:S20

Dewar AM, Haylock LA, Garner BH, Baker P, Sands RJN, Foster SP, Cox D, Mason N, Denholm I (2003) The effect of clothianidin on aphids and yellows virus in sugarbeet. Pflanzenschutz Nachrichten Bayer 56:127–146

Esker PD, Conley SP (2012) Probability of yield response and breaking even for soybean seed treatments. Crop Sci 52:351–359

Fleischer SJ, Orzolek MD, DeMackiewic D, Otjen L (1998) Imidacloprid effects on *Acalymma vittata* (Coleoptera: Chrysomelidae) and bacterial wilt in cantaloupe. J Econ Entomol 91:940–944

Glaa J, Kaiser WM (1999) Increased nitrate reductase activity in leaf tissue after application of the fungicide Kresoxim-methyl. Planta 207:442–448

Goulart ACP, Furlan SH, Fujino MT (2011) Integrated control of soybean rust (*Phakopsora pachyrhizi*) using the fungicide fluquinconazole applied as seed dressing associated with other fungicides spraying on soybean above ground parts. Summa Phytopathol 37:113–118

Gourmet C, Kolb FL, Smyth CA, Pedersen WL (1996) Use of imidacloprid as a seed-treatment insecticide to control barley yellow dwarf virus (BYDV) in oat and wheat. Plant Dis 80:136–141

Grossmann K, Kwiatkowski J, Caspar G (1999) Regulation of phytohormone levels, leaf senescence and transpiration by the strobilurin kresoxim-methyl in wheat (*Triticum aestivum*). J Plant Physiol 154:805–808

Grossmann K, Retzlaff G (1997) Bioregulatory effects of the fungicidal strobilurin Kresoxim-methyl in wheat (*Triticum aestivum*). Pesticide Sci 50:11–20

Harveson RM, Windels CE, Smith JA, Brantner JR, Cattanach AW, Giles JF, Hubbell L, Cattanach NR (2007) Fungicide registration and a small niche market: a case history of hymexazol seed treatment and the U.S. sugar beet industry. Plant Dis 91:780–790

Harvey T, Seifers DL, Kofoid KD (1996) Effect of sorghum hybrid and imidacloprid seed treatment on infestations by corn leaf aphid and greenbug (Homoptera: Aphididae) and the spread of sugarcane mosaic virus strain MDMV-B. J Agric Entomol 13:9–15

Harvey TL, Seifers DL, Martin TJ (1998) Effect of imidacloprid seed treatment on infestations of wheat curl mite (Acari: Eriophyidae) and the incidence of wheat streak mosaic virus. J Agric Entomol 15:75–81

Herms S, Seehaus K, Koehle H, Conrath U (2002) A strobilurin fungicide enhances the resistance of tobacco against *Tobacco mosaic virus* and *Pseudomonas syringae* pv. *tabaci*. Plant Physiol 130:120–127

Kuhar TP, Stivers Young LJ, Hoffman MP, Taylor AG (2002) Control of corn flea beetle and Stewart's wilt in sweet corn with imidacloprid and thiamethoxam seed treatments. Crop Prot 21:25–31

Maienfisch P, Gsell L, Rindlisbacher A (1999) Synthesis and insecticidal activity of CGA 293'343, a novel broad-spectrum insecticide. Pest Sci 55:343–389

McCornack BP, Ragsdale DW (2006) Efficacy of thiamethoxam to suppress soybean aphid populations in Minnesota soybean. Online Crop Manag. doi:10.1094/CM-2006-0915-01-RS

Milus EA, Chalkley DB (1997) Effect of previous crop, seedborne inoculum, and fungicides on development of *Stagonospora* blotch. Plant Dis 81:1279–1283

Munkvold GP, McGee DC, Iles A (1996) Effects of imidacloprid seed treatment of corn on foliar feeding and *Erwinia stewartii* transmission by the corn flea beetle. Plant Dis 80:747–749

Mutton MA, Mutton MJR, Euzebio-Filho O, Nakamura G, Aramaki P (2007) Thiamethoxam stimulates sugarcane stalk productivity. XXVI congress, international society sugar cane technology, ICC, Durban, 29 July–2 Aug 2007, pp 476–480

Nason MA, Farrar J, Bartlett D (2007) Strobilurin fungicides induce changes in photosynthetic gas exchange that do not improve water use efficiency of plants grown under conditions of water stress. Pest Manag Sci 63:1191–1200

Noleppa S, Hahn T (2013) The value of neonicotinoid insecticides in the European Union. Humboldt Forum for Food and Agriculture Working Paper 01/2013

Palumbo JC, Sanchez CA (1995) Imidacloprid does not enhance growth and yield of muskmelon in the absence of whitefly. Hortscience 30:997–999

Pataky JK, Hawk JA, Weldekidan T, Fallah MP (1995) Incidence and severity of Stewart's bacterial wilt on sequential plantings of resistant and susceptible sweet corn hybrids. Plant Dis 79:1202–1207

Pataky JK, Michener PM, Freeman ND, Weinzierl RA, Teyker RH (2000) Control of Stewart's wilt in sweet corn with seed treatment insecticides. Plant Dis 84:1104–1108

Pataky JK, Michener PM, Freeman ND, Whalen JM, Hawk JA, Weldekidan T, Teyker RH (2005) Rates of seed treatment insecticides and control of Stewart's wilt in sweet corn. Plant Dis 89:262–268

Prasanna AR, Bheemanna M, Patil BV (2004) Phytotonic and phytotoxic effects of thiamethoxam 70 WS on cotton. Karnataka J Agric Sci 17:238–241

Rice ME, Bradshaw J, Hill JH (2007) Revisiting an integrated approach to bean leaf beetle and bean pod mottle virus management. Integr Crop Manag 498:87–88

Rodriguez-Brljevich C, Kanobe C, Shanahan JF, Robertson AE (2009) Seed treatments enhance photosynthesis in maize seedlings by reducing infection with *Fusarium* spp. and consequent disease development in maize. Eur J Plant Pathol 126:343–347

Russell PE (2005) A century of fungicide evolution

Sundin DR, Bokus WW, Eversmyer MG (1999) Triazole seed treatments suppress spore production by *Puccinia recondita*, *Septoria tritici*, and *Stagonospora nodorum* from wheat leaves. Plant Dis 83:328–332

Suparyono, Pataky JK (1989) Influence of host resistance and growth stage at the time of inoculation on Stewart's wilt and Goss's wilt development and sweet corn hybrid yield. Plant Dis 73:339–345

Toquin V, Sirven C, Assmann L, Sawada H (2012) Host defense inducers 2012. In: Kramer W, Schrimer U, Witschel M (eds) Modern crop protection compounds. Wiley, Weinheim, pp 715–737

Tripathi RK, Vohra K, Schlosser E (1980) Effect of fungicides on the physiology of plants. III. Mechanism of cytokinin-like antisenescent action of carbendazim on wheat leaves. Zeitschrift Pflanzenkrankheiten Pflanzenschutz 87:631–639

Wu YX, Von Tiedemann A (2001) Physiological effects of azoxystrobin and epoxiconazole on senescence and the oxidative status of wheat. Pestic Biochem Physiol 71:1–10

Chapter 8
Non-chemical Seed Treatment in the Control of Seed-Borne Pathogens

Eckhard Koch and Steven J. Roberts

Abstract Non-chemical seed treatments include physical treatments, microbial treatments and treatments with other agents of natural origin like plant powders or extracts. Physical treatments with hot water, aerated steam, or dry heat have successfully been applied to a range of crops against a range of target pathogens and are in commercial use primarily for vegetable seeds. They can be very effective but need to be optimised on a per seed lot basis.

Microbial seed treatments may control not only seed-borne pathogens but also provide some protection against pathogenic soil-borne inoculum. However, research on the use of micro-organisms as seed treatments has been limited, and there are only a few examples of commercial use. The latter is also true for botanical seed treatments, despite many reports of bactericidal and fungicidal effects of compounds from plants. The reason may be a lack of research on the one hand but mainly commercial constraints like development and registration costs in relation to market size.

The current chapter gives an overview of approaches that have been taken to utilize the above-described non-chemical methods for control of important seed-borne pathogens of vegetables and small grain cereals. The examples treated include bacterial (black rot of brassicas, pea bacterial blight, bacterial blotch of cucurbits, black chaff of cereals), fungal (*Alternaria* diseases of carrot, black leg of brassicas, common bunt of wheat, *Fusarium* seedling diseases of small grain cereals, the loose smuts of barley and wheat, fungal diseases of rice and sorghum) and important viral diseases.

Keywords Biological control • Natural products • Organic farming • Efficacy • Bacteria • Fungi • Viruses

E. Koch (✉)
JKI, Institute for Biological Control, Heinrichstr. 243, D-64287 Darmstadt, Germany
e-mail: eckhard.koch@jki.bund.de

S.J. Roberts
Plant Health Solutions Ltd, 20 Beauchamp Road, Warwick CV34 5NU, UK
e-mail: s.roberts@planthealth.co.uk

M.L. Gullino, G. Munkvold (eds.), *Global Perspectives on the Health of Seeds and Plant Propagation Material*, Plant Pathology in the 21st Century 6,
DOI 10.1007/978-94-017-9389-6_8

1 Introduction

Seed treatment is known to have been practised since the mid seventeenth century. Hot water treatment was first reported in the 1880s, but for most of the history of modern agriculture, the majority of seed treatments have been chemicals targeting fungal pathogens, with physical treatments used for bacteria and viruses where there were no chemical options. The growing public concern about environmental risks associated with the use of agrochemicals, the political will to reduce pesticide use (as in the EU), and the development of the organic movement that prohibits the use of synthetic seed treatments all contribute to the current resurgence of interest in non-chemical seed treatments. To many people "non-chemical" means something coming from nature that is safe for humans and the environment. In the context of plant protection and IPM the term generally encompasses precautionary measures, the utilisation of natural mechanisms of control as well as treatments with control agents that are not chemically synthesised. For the purposes of this chapter we follow the latter concept and define non-chemical seed treatments as including all those treatments which are not considered as conventional synthetic pesticides. Thus we will consider physical treatments, microbial treatments, and treatments with natural products.

The main requirements for an "ideal" seed treatment are identical for chemical and non-chemical seed treatment methods. Both should in the first place reduce the numbers or transmission of the target pathogen(s) from the seed to the shoot to acceptable levels. They should further: not reduce germination or vigour; not reduce storability; have low toxicity to humans/animals; not harm the environment. For any kind of seed treatment the location of the pathogen on the seed has significant implications for the likelihood of achieving satisfactory levels of control. Pathogen inoculum may be superficial or internal. Superficial inoculum is located on the surface of the seed/fruit (most bacteria and many fungi) and is easier to eradicate. Internal inoculum may be located in the testa/pericarp (many fungi, some viruses), in the endosperm/cotyledons (a few fungi), or in the embryonic axis (viruses, certain smuts). Chemical seed treatments may contain different active ingredients which may also protect against soil-borne pathogens or, in the case of systemic compounds, provide transient protection against air-borne inoculum, e.g. from powdery mildew. Non-chemical seed treatments have activity primarily against pathogens on or in seeds, some may in addition provide a certain level of protection against soil-borne pathogens.

There is a lack of consistency amongst the various studies in the way results are interpreted and summarized. It is important to consider that the efficacy of seed treatments can only be determined in terms of the seed test or trials used to assess them. Thus it is vital to pay attention to the details of the tests or trials used to assess treatments, particularly the numbers of seeds examined or sown. We have therefore attempted to introduce some consistency by careful re-examination of results in terms of the detection limit or tolerance standard (Roberts et al. 1993) of the test applied. Where possible, we indicate efficacy in terms of the percentage reduction in seed infestation levels or pathogen numbers achieved by the treatment.

In this chapter we summarise the principle non-chemical seed treatment methods, give an overview of approaches that have been taken and also include information on non-chemical seed treatment products and technologies that are already in commercial use. In view of the extensive literature and the large number of seed-borne pathogens we will give specific examples for some important crops. We will preferably summarize results from field- and greenhouse experiments, where available, and avoid laboratory results e.g. on in-vitro testing aimed at characterising the fungicidal or bactericidal potential of putative control agents. The goal is to provide an overview of the current status of non-chemical seed treatments. The information should allow the identification not only of bottlenecks but also their future potential and prospects.

2 Principal Methods of Non-chemical Seed Treatment

2.1 *Physical Treatments*

Physical treatments have a number of advantages over other treatments: in most countries they do not require registration or approval; they have a wide spectrum of activity; they do not leave any toxic or polluting residues. The latter means that treated seed can also be used for other purposes, e.g. animal feed. The main disadvantages are: the need for optimisation on a per seed lot basis; possible high energy and capital costs; no effects on soil-borne pathogens.

2.1.1 Heat Treatments

Heat treatment or thermotherapy is based on the principal that pathogens are often killed or inhibited at temperatures that are not, or only slightly, deleterious to the seed (Baker 1962). Due to differences in thermal exchange efficiencies, the temperature or time required for successful treatment increases in the order: hot water, aerated steam, dry air (Baker 1972).

Before effective chemical seed treatments became available in the second half of the twentieth century, hot water treatments were widely used for the sanitization of vegetable and cereal seeds. The main disadvantage is the need for post treatment drying. Hot water treatments on cereals (e.g. for loose smut control) and vegetables (e.g. for control of *Phoma lingam* or *Xanthomonas campestris* on crucifer seeds) were commonly performed at temperatures between 50 °C and 55 °C and with durations from 3 to 25 min (Walker 1948; Baker 1962; Gratwick and Southey 1986). A compilation of hot water seed treatment conditions for different vegetables is available on the internet (McGrath 2013). For control of specific cereal pathogens, certain variants were in use, like a discontinuous hot water treatment or a warm water treatment (Jahn 2008). Hot water treatments are still important for the treatment of various kinds of vegetative plant propagation material. Examples of commercial use include the eradication of the bacterium *Leifsonia xyli* subsp. *xyli*, the causal agent of ratoon stunting disease of sugarcane from seed canes (Johnson and Tyagi 2010), the management of nematodes transmitted by suckers of banana and plantain (Coyne et al. 2010) and nematodes in narcissus bulbs (Qiu et al. 1993). For general descriptions of the treatment of vegetative plant propagation material by physical methods the reader is referred to the overviews by Baker (1962), Gratwick and Southey (1986) and Grondeau and Samson (1994).

Compared to treatment in water, the main advantages of seed treatment with aerated steam are a more accurate temperature control, usually less impairment of seed germination and that the seeds are left much dryer. On the other hand, there has been only limited success against bacterial diseases (Baker 1972; Navaratnam et al. 1980). In Sweden, a technology has been developed that is based on high precision control of treatment temperature and humidity and application of the aerated steam in fluidized beds (Thermoseed®) (Forsberg et al. 2005). Over the last decade, high throughput devices (0.2–15 tons per hour) have been constructed and are in commercial use in Sweden, Norway and the Netherlands for the treatment of cereal and vegetable seeds (G. Forsberg, pers. communication).

Due to the comparatively long treatment durations required for pathogen inactivation (from a few days to 2 weeks or longer), seed treatments with dry heat often cause reductions in seed germination. However, they do not require sophisticated equipment and are therefore easy to apply. There are relatively few reports that claim successful control of seed-borne bacteria (e.g. Kubota et al. 2012) or fungi (e.g. Gilbert et al. 2005) by dry heat. In contrast, inactivation of viruses, both in vegetative propagation material and in seeds by dry heat treatments is well documented (Nyland and Goheen 1969; Grondeau and Samson 1994).

2.1.2 Other Physical Treatments

A seed treatment technology based on the application of low energy electrons (e-ventus®) has been developed in Germany. It is mainly effective against pathogens on the seed surface, like the spores of common bunt (*Tilletia caries*) or rye stripe smut (*Urocystis occulta*) (Jahn et al. 2005), and has also shown activity against a number of seed-borne vegetable pathogens. Various other physical effects such as high frequency fields, ultrasonic waves or microwaves have been studied for their suitability as seed treatments (Baker 1972; Bhaskara Reddy et al. 1998) but so far not been successful enough to be commercialized.

2.2 *Micro-organisms and Natural Products*

2.2.1 Micro-organisms

The basic mechanisms underlying biological control of plant pathogens are hyperparasitism, suppression by antibiotics, lytic enzymes or other metabolites, and competitive exclusion. Micro-organisms may also elicit host defences; strains of root-colonizing bacteria have been identified as potential elicitors of plant host defences. In several instances, inoculation with plant-growth-promoting rhizobacteria (PGPR) resulted in control of multiple diseases caused by different pathogens (Pal and McSpadden Gardener 2006). The majority of micro-organisms used as biocontrol agents originate from plants, especially from the rhizoplane or from the rhizosphere. In recent studies it has been shown that induction of resistance and increased stress tolerance can also be triggered by seed-application of bacterial endophytes, i.e. strains originating from the interior of plants (Joe et al. 2012; Fürnkranz et al. 2012). For marketing as seeds treatment, the micro-organisms must be formulated in an appropriate way to ensure efficacy, storability and compatibility with existing agricultural technologies and practises. One way of delivery of micro-organisms to vegetable seed is by adding them during the priming process ('biopriming') (Jensen et al. 2007; Bennett et al. 2009; Pill et al. 2009). Microbial inocula or other natural products are not only added to seeds for technical reasons like plant growth promotion or disease control. They also provide the seed with a "green" label that is used in marketing.

2.2.2 Plant-Derived Products

Plants are a relatively untapped reservoir of different chemicals that can be used directly or serve as templates for the development of pesticides (Yoo et al. 2013). There is a large body of literature describing plants or plants constituents with antimicrobial properties. Activity against bacterial plant pathogens has been reported particularly for essentials oils (Iacobellis et al. 2005; van der Wolf et al. 2008; Mengulluoglu and Soylu 2012). Fungicidal activity of plant extracts has been shown against a large number of seed-borne fungi including members of important genera such as, e.g., *Fusarium*, *Alternaria* or *Colletotrichum* (Dal Bello and Sisterna 2010; Marinelli et al. 2012). The plants or plant parts may be used as powders or as extracts obtained by water or solvent extraction. However, the use of plant extracts in plant protection is often limited or infeasible due to phytotoxic properties of the preparations. Due to the general high sensitivity of germinating seed to external stimuli this holds especially true for the use as seed treatments. A typical example are the plant essential oils whose strong antimicrobial activity is often associated with adverse effects on the seed germination process (Dudai et al. 1999; Tworkoski 2002).

3 Examples of Non-chemical Seed Treatments for Control of Bacterial Pathogens

Due to a lack of chemical options, there has been more focus on non-chemical treatments for the control of seed-borne bacteria over recent years than for fungi. As bacterial pathogens are more important on vegetable crops than cereals there has inevitably been more work on these crops. Generally also, because of the great potential for secondary spread under favourable conditions, especially in transplanted vegetables (Roberts et al. 1999; Roberts et al. 2007), the seed health standards that need to be achieved are higher than for many fungal diseases.

3.1 Black Rot of Brassicas

Black rot of brassicas, caused by *Xanthomonas campestris* pv. *campestris*, is probably one of the most important diseases of brassicas worldwide. Hot water treatment has long been used as a seed treatment (Clayton 1924), but can result in reduced germination (Huber and Gould 1949). It can be very effective at reducing inoculum levels when optimised on a per seed lot basis. In recent experiments (Roberts et al. 2006), hot water and aerated steam consistently reduced seed infestation levels and seed-to-seedling transmission. Reductions in seed infestation of over 90 % and in transmission of over 63 % were achieved. Electron treatment was also examined, but this was less consistent and effective than hot water or aerated steam. Reductions in transmission have also been demonstrated for several microbial treatments; these have included experimental micro-organisms and commercial Bacillus products, Serenade and Subtilex (Roberts 2009) (Fig. 8.1). Thyme oil has also been shown to have potential as seed treatment for *X. campestris* pv. *campestris* on brassicas (van der Wolf et al. 2008; Roberts 2009) (Fig. 8.1).

3.2 Pea Bacterial Blight

Pea bacterial blight is caused by *Pseudomonas syringae* pv. *pisi*, and can cause significant losses particularly in over-wintered crops. The disease is primarily seed-borne and the use of disease-free seed is the main means of control (Roberts et al. 1996). Grondeau et al. (1992) examined a range of heat treatments in the form of hot water, hot humid air and dry heat treatments. Moist heat (50 °C, 100 % humidity, 48 h) reduced germination to un-acceptable levels and was not pursued further. Hot water (15 min, 55 °C) gave at least 75–84 % reductions in the percentage of seeds infested, and dry heat (65 °C, 72 h) gave 66–83 % reductions, without major effects on germination.

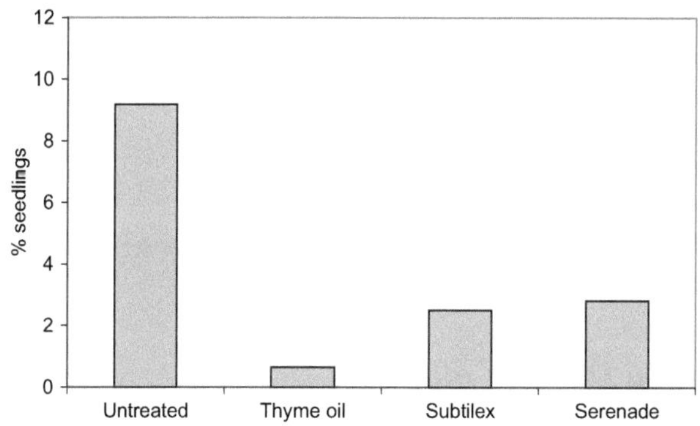

Fig. 8.1 Mean transmission of *Xanthomonas campestris* pv. *campestris* in three brassica seedlots grown as module transplants in the glasshouse, following treatment of the seed with biologicals/ natural products. Values are the mean % of seedlings infested (i.e. contaminated or infected) (From Roberts 2009)

3.3 Bacterial Blotch of Cucurbits

Bacterial blotch of cucurbits is caused by *Acidovorax citrulli*, and can result in total crop loss in water melon crops (Latin and Hopkins 1995).

Rane and Latin (1992) obtained 80 % (naturally infested seed) and 96 % (laboratory infested seed) reductions in seed transmission with hot water treatment (50 °C, 20 min). Kubota et al. (2012) using dry heat (85 °C, 3–5 days) claimed complete disinfection. However, the maximum number of seeds tested was 300, implying tolerance standard of 1 % (see Roberts et al. 1999). Thus, with an initial infestation level of 25 %, they actually achieved a 96 % or greater reduction for melon and 97 % reduction for cucumber. As with dry heat for control of Tobamoviruses in tomatoes (see below), pre-drying seed to low moisture contents <5 % seems to reduce the likelihood of damage during heat treatment.

The use of a non-pathogenic (genetically modified) *A. citrulli* strain has been examined as a potential seed treatment (Johnson et al. 2011). They achieved an 82 % reduction in disease in growth chamber tests, but only a 38 % reduction in glasshouse tests. Seed treatment with a cell-free culture filtrate of a yeast has also resulted in a significant reduction in disease (Wang et al. 2009).

3.4 Black Chaff of Cereals

Black chaff and bacterial leaf stripe of cereals (barley, rye, wheat and triticale) are caused by the seed-borne bacterium *Xanthomonas translucens* pv. *translucens*. It has been reported from all continents, but is listed as a quarantine pathogen in some regions/countries. Outbreaks of the disease are sporadic and favoured by warm and moist conditions. Treatment of barley seed with dry heat (71–84 °C, 11 days) reduced bacterial numbers to undetectable levels from an initial level of over

10^6 CFU/g (Fourest et al. 1990). For a seed lot with a lower initial level, 4 days at 72 °C was effective. The authors recommended routine treatment at 72 °C for 5–7 days as the higher infestation level is unusual. However, Duveiller et al. (1997) comment that the method is not completely effective.

4 Examples of Non-chemical Seed Treatments for Control of Fungal Pathogens

4.1 Alternaria Diseases of Carrot

Leaf blight and black root rot of carrots are caused by *Alternaria dauci* and *A. radicina*. Both pathogens are seed-borne and also contribute to poor emergence. In an extensive study of non-chemical treatments for these pathogens, the efficacies of physical, microbial and natural products were compared (Koch et al. 2010). Treatments were evaluated in both controlled conditions and in field trials over several years, mainly on the basis of plant stand. A number of putative resistance inducers failed to give any control. In five field trials performed in four different countries significant improvements were obtained with hot water, aerated steam and electron treatment, two microbial treatments and an emulsion of thyme oil in water. The most effective treatments (hot water plus *C. rosea* IK726 and aerated steam treatment) resulted in almost a 100 % increase in plant stand (Fig. 8.2).

4.1.1 Black Leg of Brassicas

Phoma lingam (*Leptosphaeria maculans*) causes black leg and stem canker of brassicas. It can also cause death and damping-off of seedlings. Williams (1967) found that hot water treatment (50 °C, 25 min) reduced infestation levels by 89 %, but this was not adequate to prevent a serious outbreak of the disease for a seed lot with high levels of infestation (18 %). A number of non-chemical (microbial) treatments and hot water have also been investigated more recently (Clarkson and Roberts 2011). The best treatments: hot water (50 °C, 30 min), thyme oil, Serenade (*Bacillus subtilis*) and an experimental microbial product significantly reduced seed infestation levels compared to the untreated control and were as effective as thiram. Hot water treatment was the most effective and resulted in an 88 % reduction in transmission. However, it also resulted in a significant reduction in emergence and an increase in damping-off caused by *Pythium* spp.

4.2 Common Bunt of Wheat

Common bunt of wheat is caused by *Tilletia caries* and *T. laevis*. Individual grains are replaced by masses of black spores which are dispersed to healthy grains at harvest and during grain handling. For effective control, high seed health standards

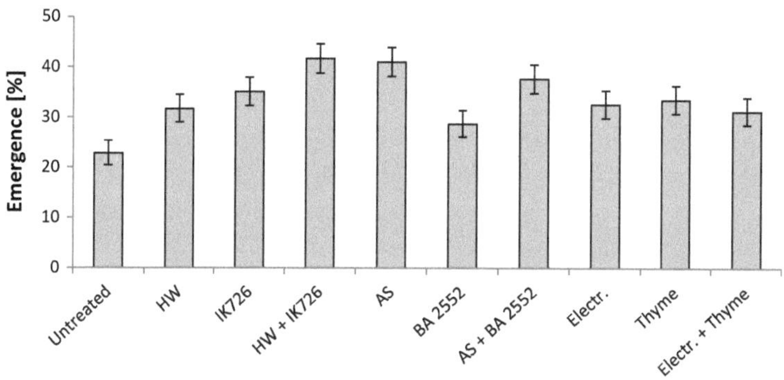

Fig. 8.2 Effect of selected seed treatments or treatment combinations on establishment of carrot plants developing from seeds naturally infected with *Alternaria dauci* and *A. radicina*. Means of five field experiments performed in 2006 in Sweden, UK, Italy and Germany. Error bars show approximate 95% confidence intervals; means with non-overlapping confidence intervals were considered to be significantly different (*HW* hot water; IK726: *Clonostachys rosea* IK726; *AS* aerated steam; BA2552: experimental formulation of *Pseudomonas chlororaphis* MA342; *Electr* electron treatment; Thyme: Emulsion of thyme oil in water) (Koch et al. 2010)

are considered necessary. Thresholds for spore load recommended in different countries vary between <1 and 20 spores per seed and should be adapted to the susceptibility of the variety (Waldow and Jahn 2007).

Seed treatment with different organic substances such as skimmed milk powder or wheat flour has been shown to be effective experimentally (Becker and Weltzien 1993). However, one of the obstacles for commercialization is that technologies such as seed pelleting are required to apply the needed large amounts of material to the seed. A product based on yellow mustard powder (Tillecur®) is commercialized in several European countries where it is used primarily for bunt control in organic farming (Waldow and Jahn 2007).

Numerous experimental and commercialized micro-organisms have been examined as seed treatments on wheat against common bunt (Hökeberg et al. 1997; Koch et al. 2004; Koch et al. 2006; Goates and Mercier 2011). However, so far only the bacterium *Pseudomonas chlororaphis* MA342 has been commercialised and is marketed in Europe as an oil-based formulation (Cedomon®) for hulled seeds (like barley) and as a water-based formulation (Cerall®) for non-hulled seeds (like wheat). In field trials with spelt (*Triticum spelta*), treatment with Cedomon® reduced disease incidence by almost 90 % and with Cerall® on dehulled spelt an 80 % reduction in disease incidence was recorded (Krebs 2010).

Control of common bunt with levels almost equivalent to chemical seed treatments haves been reported for both electron treatment (mean reductions of 87–94 %) (Jahn et al. 2005) and aerated steam (95 % reduction) (Forsberg et al. 2005).

4.3 Fusarium *spp. and* Microdochium *spp. on Small-Grain Cereals*

Fungal pathogens belonging to species of *Fusarium*, *Microdochium*, *Phaeosphaeria*, *Pyrenophora* and *Rhynchosporium* affect primarily germination and seedling health of small grain cereals. They are mostly located in the pericarp (outer layer of the grain). In the following, options for non-chemical control of this group of seed-borne pathogens will be explained using the example of *Fusarium* spp. and *Microdochium* spp. Both cause reductions in germination and seedling losses. Under a snow cover, *Microdochium majus* and *M. nivale* may cause snow mould. *Fusarium* spp. and *Microdochium* spp. may under favourable conditions also penetrate deeper into the endosperm or even colonise the embryo.

Seed infections with *Fusarium* spp. and *Microdochium* spp. have been successfully controlled with warm water (45 °C, 2 h; increase in plants/m row by 160 %; Vogelgsang 2013). Positive effects were also obtained with aerated steam treatments (increase in crop density of 20 %, chemical: 28 %; Forsberg et al. 2005). Treatment with dry heat at 70 °C for 5 days was recommended for eradicating *F. graminearum* from wheat seeds (Clear et al. 2002; Gilbert et al. 2005). Microwave irradiation has also been shown to reduce the percentage of wheat seed infected with *Fusarium graminearum* by at least 74 % (Bhaskara Reddy et al. 1998).

Seed treatment with fungal antagonists belonging to species of *Trichoderma*, *Gliocladium* and *Penicillium* were reported to reduce foot and root rot caused by *F. culmorum* in field experiments in Italy, although to a lesser extent than the chemical (Roberti et al. 2000). In seed tray tests with wheat seed lots naturally infected with *Fusarium* spp. significant increases in the number of healthy seedlings were obtained with *Streptomyces antimycoticus* strain FZB53 (Koch et al. 2006). Analytical studies indicated that the activity was largely due to an unidentified polyether antibiotic and geldanamycin produced by the antagonist (Koch et al. 2008). In field and greenhouse experiments using seeds artificially inoculated with *F. culmorum* disease indices on seedlings of barley and wheat were repeatedly reduced by more than 80 % by seed treatment with *Clonostachys rosea* IK726 (syn. *Gliocladium roseum*) (Jensen et al. 2000). Chitosan, a polymer of β-1,4 linked D-glucosamine applied to wheat seed infected with *F. graminearum* significantly improved seed germination and at the higher concentrations tested inhibited fungal transmission to the primary roots of germinating seedlings by >50 %. The observed effects were attributed to activation of plant defence, although a partial contribution of the antifungal properties of chitosan could not be totally ruled out (Bhaskara Reddy et al. 1999).

Preparations from Chinese galls (obtained from *Rhus chinensis*) have recently been shown to have potential to control seed-borne *Microdochium majus* (Vogelgsang et al. 2013). In vitro, Chinese galls at a concentration of 1 % inhibited *M. majus* conidial germination almost as effectively as the synthetic fungicide Pronto® Plus (spiroxamine + tebuconazole). In field experiments, over three

Fig. 8.3 Emergence of wheat from kernels naturally infected with *Microdochium majus*. *Left*: Untreated control, *Right*: seed treated with Chinese galls + adhesive (Courtesy of H. Krebs, Agroscope, Zürich)

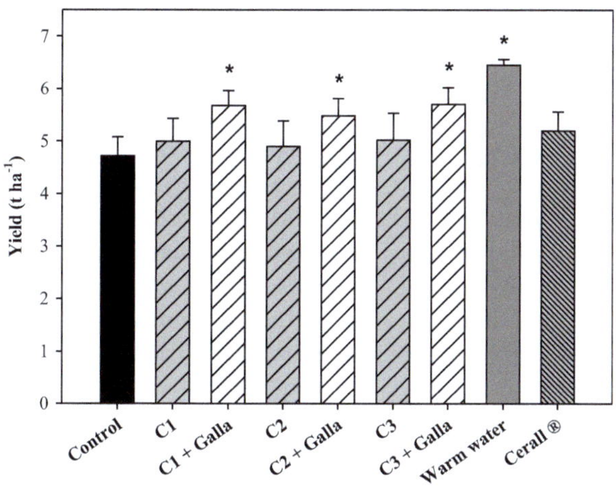

Fig. 8.4 Effect of seed treatment of wheat infected with *Microdochium majus* on grain yield in field experiments 2009–2011. C1, C2, C3: different application procedures of the adhesive; Galla: *Rhus chinensis* (2 g per 100 g seeds); warm water: 45 °C, 2 h; Cerall: 1 ml per 100 g seeds. Means and standard errors of means. *Asterisks* indicate significant differences (α=0.05) to the control. (Vogelgsang et al. 2013, with permission)

years, treatment of infected wheat kernels with three different formulations of Chinese galls resulted in significant increases in emergence (Fig. 8.3) and yield (Fig. 8.4). Chinese galls are known to contain tannin-derived components with low pH, but whether these have a role in the antifungal activity is not conclusive (Vogelgsang et al. 2013).

4.4 *Loose Smuts of Barley and Wheat*

Due to their localization in the seed embryo and early colonization of the apical meristem (Wunderle et al. 2012), the loose smut fungi *U. nuda* and *U. tritici* are particularly difficult to control. To the authors' knowledge, effective sanitization of infected seed lots by non-chemical methods is only possible by thermal treatment in water. The effect obtained with aerated steam was only partial and clearly lower than with the standard chemical seed treatment (Forsberg et al. 2005). A range of plant extracts and microbial antagonists with in-vitro activity against germinating spores of *U. nuda* have been screened in field trials, but none gave satisfactory control (Koch, unpublished results).

4.5 Diseases of Sorghum

In Nigeria, Ghana and Burkina Faso *Fusarium*, *Curvularia* and *Phoma* are common on sorghum (*Sorghum bicolor*) seed; they affect seed germination and seedling health (Zida et al. 2012). *Phoma sorghina* (teleomorph *Leptosphaeria sacchari*) is primarily located in the seed coat but can also be found in the endosperm and embryo, although at lower frequency (Schémaeza et al. 2012). Among different aqueous plant extracts tested for activity against *P. sorghina*, the most effective were those from *Cymbopogon citratus* (30 % W/V, treatment duration 24 h) and *Eclipta alba* (10 % W/V, treatment duration 20 h). An aqueous extract of *Yucca schidigera* showed antifungal activity against *P. sorghina*, *Fusarium* spp., *Cochliobolus lunatus* and *Cladosporium* spp. and increased seedling emergence and seedling vigour. The activity was suspected to be due to saponins present in the extract (Wulff et al. 2012).

Similarly, in field experiments using inoculated seeds, treatment with dried powder from the berries of African soapberry (*Phytolacca dodecandra*), known to contain saponins, reduced the disease incidence of covered kernel smut (*Sporisorium sorghi*) (Fig. 8.5) and loose kernel smut (*S. cruentum*) by 82–92 % (Tegegne and Pretorius 2007). A crude extract from aerial parts of *Agapanthus africanus* controlled both smuts completely (Tegegne et al. 2008). Results of seed treatment experiments performed in the glasshouse indicated a high activity against *S. sorghi* also for Tillecur®, an aqueous extract of *Quillaja saponaria* and *Trichoderma harzianum* (Moharam 2010).

4.6 Diseases of Rice

A large number of plant extracts have been screened for fungicidal activity against rice pathogens (e.g. Mohana et al. 2011), but few studies involved testing on infected seed. Garlic extracts applied to rice seeds were as effective at reducing the incidence of different seed-borne pathogens as the synthetic chemical reference fungicide (Yeasmin et al. 2012). Broad spectrum fungicidal activity was also recorded for the essential oils of *Cymbopogon citratus*, *Ocimum gratissimum* and *Thymus vulgaris* applied as emulsions in 0.1 % agar to rice seeds. The treatments

Fig. 8.5 Sorghum panicle with healthy seeds and sori of covered kernel smut. The sori are crushed at harvest and adhere to the seed surface from where they infect the germinating plant

reduced seed infection with *Alternaria padwickii*, *Bipolaris oryzae* and *Fusarium moniliforme* in blotter tests and seed to seedling transmission in pot experiments by 76–95 % (Nguefack et al. 2008).

In Japan, *Trichoderma asperellum* SKT-1 (Ecohope®) and *Talaromyces flavus* SAY-Y-9401 (Tough-block®) are registered as seed treatments for control of seed-borne *G. fujikuroi* (Nagayama et al. 2007; Kato et al. 2012) and other seed-borne pathogens of rice.

5 Examples for Non-chemical Seed Treatments for Control of Viruses

There are relatively few examples of non-chemical seed treatments for viruses. Probably this is because, for many viruses, the virus must be present in the embryonic axis for transmission to occur, and presents a difficult target for treatment without damaging the seed.

5.1 Solanacae and Tobamovirus

The tobamoviruses TMV (tobacco mosaic virus) and ToMV (tomato mosaic virus) are mechanically transmissible and able to retain their infectivity in the seed coat of dry tomato seeds. Dry heat (80 °C, 24 h) reduced the transmission of TMV in tomato to undetectable levels in most, but not all, seed lots (Laterrot and Pécaut 1968). In the one seed lot where transmission was detected, a reduction of 70 % was achieved. Dry heat treatment (78 °C, 2 days) reduced the levels of ToMV in tomato seed by over 95 % (based on the number of local lesions in a host test) without detrimental effects on germination after storage for 12 months (Green et al. 1987). Prior to heat treatment seeds were brought to a moisture content of between 6 % and 8 %. Dry heat (80 °C, 24 h) has also been shown to reduce levels of ToMV in pepino seeds to undetectable levels (<3 %) but led to a reduction in germination that varied between species (Prohens et al. 1999).

5.2 *Melon Necrotic Spot Virus*

Melon necrotic spot virus is an important pathogen of glasshouse and field-grown melons and cucumbers. The effect of dry heat at 70 °C for 3–6 days on germination and virus transmission was examined by Herrera-Vásquez et al. (2009). The best treatment (6 days at 70 °C) gave at least 80–86 % reductions in transmission (although the authors interpreted this as total eradication) with little effect on germination.

6 Summary

There are a number of non-chemical seed treatment options for the control of seed-borne diseases. No treatment can be guaranteed to completely eliminate the target pathogen, and claims of 'eradication' or 'complete control' should be regarded with some suspicion. Physical treatments with hot water, aerated steam, or dry heat have successfully been applied to a range of crops against a range of target pathogens and are in commercial use. The level of success achieved depends on the location of inoculum and optimisation of the treatment parameters for different species and seed lots. Hot water and aerated steam seem to perform relatively better on small seeds when the inoculum is mostly superficial, whereas dry heat may be better for viruses, with larger seeds and/or where the inoculum is more deep-seated. Where feasible, the efficacy of treatment should be checked with a post-treatment seed test. An advantage of the physical treatments is that they can target a number of pathogens at the same time, but the non-specific nature and potential for sub-lethal damage of seed can potentially give rise to other problems such as increased susceptibility to soil-borne pathogens. Expansion of commercial use of the physical treatments has been hampered by a number of factors such as: problems associated with the batch treatment of large bulks of seeds, the difficulty of applying precise treatments, the need for re-drying after hot water treatment, a lack of commercial equipment, and the need for optimisation of treatment parameters for each seed lot. Recently-developed innovative technologies (Thermoseed®, e-ventus®) are overcoming some, but not all, of these obstacles.

Microbial seed treatments have had much more variable levels of success. It is interesting to note that many of the microbial antagonists that protect against root or foliar pathogens were reported to have been applied by seed treatment. The literature specifically describing control of seed-borne diseases by microbial antagonists is nevertheless limited compared to the huge number of reports on microbial control of other kinds of plant diseases. Only a few microbial seed treatments for control of seed-borne pathogens are commercially available. This may be due to a lack of research on the one hand but more often due to commercial constraints like development costs in relation to market size, the feasibility of mass-production, formulation, and general difficulties associated with the registration of microbials as plant protection products. Also, where microbial treatments have been originally developed for foliar application, there seems little incentive to seek approval as a seed treatment, as the potential market volume is much lower. However, with changes to legislation in the EU and following several recent take-overs of smaller biocontrol-focused companies by large multi-nationals, it is possible that the rate of progress will increase over the next few years.

Despite many reports of bactericidal and fungicidal effects of compounds from plants in the literature, the use of botanicals as seed treatments is still rare. As with microbial products, economic considerations like cost of registration and limited attractiveness for the market are likely to be the main reasons. Also, companies find it difficult to claim intellectual rights for products from plants with published antimicrobial properties. An alternative to commercialization of botanical seed treatments by companies could be self-preparation by the user, provided this would be in line with legislation (for example use of plant material as basic substance according to regulation EC No. 1107/2009). The use of non-chemical seed treatments based on natural products such as powders or extracts from local plants could also be a sustainable solution particularly for many developing countries where chemical seed treatments are unaffordable or not available to the farmer.

Finally, we conclude that for many seed-borne pathogens, there are potentially effective non-chemical seed treatment alternatives to synthetic chemicals available or that could be developed. The fact that they haven't been exploited more fully seems to be largely due to a lack of commercial incentives; this may change with the increasing concerns about the safety of synthetic chemical pesticides.

References

Baker KF (1962) Thermotherapy of planting material. Phytopathology 52:1244–1255

Baker KF (1972) Seed pathology. In: Kozlowski TT (ed) Seed biology, vol 3. Academic, New York/London, pp 318–416

Becker J, Weltzien HC (1993) Bekämpfung des Weizensteinbrandes (*Tilletia caries* (D.C.) Tul. & C. Tul.) mit organischen Nährstoffen. Z Pflanzenkrankh Pflanzenschutz 100:49–57

Bennett AJ, Mead A, Whipps JM (2009) Performance of carrot and onion seed primed with beneficial microorganisms in glasshouse and field trials. Biol Control 51:417–426

Bhaskara Reddy MV, Raghavan GSV, Kushalappa AC, Paulitz TC (1998) Effect of microwave treatment on quality of wheat seeds infected with *Fusarium graminearum*. J Agric Eng Res 71:113–117

Bhaskara Reddy MV, Arul J, Angers P, Couture L (1999) Chitosan treatment of wheat seeds induces resistance to *Fusarium graminearum* and improves seed quality. J Agric Food Chem 47:1208–1216

Clarkson JC, Roberts SJ (2011) Disease management in organic brassica seed and transplants HDC project FV 352, final report 2010-11. AHDB, Stoneleigh. http://www.hdc.org.uk/sites/default/files/research_papers/FV%20352%20Final%20report%202011_0.pdf

Clayton EE (1924) Investigations of cauliflower diseases on Long Island. N Y State Agric Exp Stat Bull 506:31

Clear RM, Patrick SK, Turkington TK, Wallis R (2002) Effect of try heat treatment on seed-borne *Fusarium graminearum* and other cereals. Can J Plant Pathol 35:489–498

Coyne D, Wasukira A, Dusabe J, Rotifa I, Dubois T (2010) Boiling water treatment: A simple, rapid and effective technique for nematode and banana weevil management in banana and plantain (*Musa* spp.) planting material. Crop Prot 29:1478–1482

Dal Bello G, Sisterna M (2010) Use of plant extracts as natural fungicides in the management of seedborne diseases. In: Arya A, Perelló AE (eds) Management of fungal plant pathogens. CAB International, Wallingford, pp 51–66

Dudai N, Poljakoff-Mayber A, Mayer AM, Putievsky E, Lerner HR (1999) Essential oils as allelochemicals and their potential use as bioherbicides. J Chem Ecol 25:1079–1089

Duveiller E, Fucikovsky L, Rudolph K (eds) (1997) The bacterial diseases of wheat: concepts and methods of disease management. CIMMYT, Mexico

Forsberg G, Johnsson L, Lagerholm J (2005) Effects of aerated steam treatment on cereal diseases and crop yield. J Plant Dis Prot 112:247–256

Fürnkranz M, Adam E, Müller H, Grube M, Huss H, Winkler J, Berg G (2012) Promotion of growth, health and stress tolerance of styrian oil pumpkins by bacterial endophytes. Eur J Plant Pathol 134:509–519

Fourest E, Rehms LD, Sands DC, Bjarko M, Lund RE (1990) Eradication of *Xanthomonas campestris* pv. *translucens* from barley seed with dry heat treatments. Plant Dis 74:816–818

Gilbert J, Woods SM, Turkington TK, Tekauz A (2005) Effect of heat treatment to control *Fusarium graminearum* in wheat seed. Can J Plant Pathol 27:448–452

Goates BJ, Mercier J (2011) Control of common bunt of wheat under field conditions with the biofumigant fungus *Muscodor albus*. Eur J Plant Pathol 131:403–407

Gratwick M, Southey JF (eds) (1986) Hot-water treatment of plant material. Reference book 201. Her Majesty's Stationery Office, London

Green SK, Hwang LL, Kuo YJ (1987) Epidemiology of tomato mosaic virus in Taiwan and identification of strains. Z Pflanzenkrankh Pflanzenschutz 94:386–397

Grondeau C, Ladonne F, Fourmond A, Poutier F, Samson R (1992) Attempt to eradicate *Pseudomonas syringae* pv. *pisi* from pea seeds with heat treatments. Seed Sci Technol 20:515–525

Grondeau C, Samson R (1994) A review of thermotherapy to free plant material from pathogens, especially seeds from bacteria. CRC Crit Rev Plant Sci 13:57–75

Herrera-Vásquez JA, Córdoba-Sellés MC, Cebrián MC, Alfaro-Fernández A, Jordá C (2009) Seed transmission of Melon necrotic spot virus and efficacy of seed-disinfection treatments. Plant Pathol 58:436–442

Hökeberg M, Gerhardson B, Johnsson L (1997) Biological control of cereal seed-borne diseases by seed bacterization with greenhouse-selected bacteria. Eur J Plant Pathol 103:25–33

Huber GA, Gould CJ (1949) Cabbage seed treatment. Phytopathology 39:869–874

Iacobellis NS, Lo Cantore P, Capasso F, Senatore F (2005) Antibacterial activity of *Cuminum cyminum* L. and *Carum carvi* L. essential oils. J Agric Food Chem 53:57–61

Jahn M, Röder O, Tigges J (2005) Electron treatment of cereal crop seeds – overview and appraisal of field trials. Mitteilungen aus der Biologischen Bundesanstalt für Land- und Forstwirtschaft Berlin-Dahlem 399:66–128

Jahn M (2008) Physikalische Beizung. In: Kruse M (ed) Handbuch Saatgutaufbereitung. AgriMedia, Clenze, pp 161–173

Jensen B, Knudsen IMB, Jensen DF (2000) Biological seed treatment of cereals with fresh and long-term stored formulations of *Clonostachys rosea*: Biocontrol efficacy against *Fusarium culmorum*. Eur J Plant Pathol 106:233–243

Jensen DF, Knudsen IMB, Lübeck M, Mamarabadi M, Hockenhull J, Jensen B (2007) Development of a biocontrol agent for plant disease control with special emphasis on the near commercial fungal antagonist *Clonostachys rose*a strain 'IK726'. Australasian Plant Pathol 36:95–101

Joe MM, Islam MR, Karthikeyan B, Bradeepa K, Sivakumaar PK, Sa T (2012) Resistance responses of rice to rice blast fungus after seed treatment with the endophytic *Achromobacter xylosoxidan*s AUM54 strains. Crop Prot 42:141–148

Johnson SS, Tyagi AP (2010) Effect of ratoon stunting disease (RSD) on sugarcane yield in Fiji. South Pacific J Nat Appl Sci 28:69–73

Johnson KL, Minsavage GV, Le T, Jones JB, Walcott RR (2011) Efficacy of a nonpathogenic *Acidovorax citrull*i strain as a biocontrol seed treatment for bacterial fruit blotch of cucurbits. Plant Dis 95:697–704

Kato A, Miyake T, Nishigata K, Tateishi H, Teraoka T, Arie T (2012) Use of fluorescent proteins to visualize interactions between the Bakanae disease pathogen *Gibberella fujikuroi* and the biocontrol agent *Talaromyces* sp. KNB-422. J Gen Plant Pathol 78:54–61

Koch E, Schmitt A, Stephan D, Kromphardt C, Jahn M, Krauthausen HJ, Forsberg G, Werner S, Amein T, Wright SAI, Tinivella F, Gullino ML, Roberts SJ, van der Wolf J, Groot SPC (2010) Evaluation of non-chemical seed treatment methods for the control of *Alternaria dauci* and *A. radicina* on carrot seeds. Eur J Plant Pathol 127:99–112

Koch E, Weil B, Eibel P (2004) Development of a leaf symptom-based screening method for seed treatments with activity against *T. caries* and application of the method using microbial antagonists. J Plant Dis Prot 111:470–483

Koch E, Weil B, Wächter R, Wohlleben S, Spiess H, Krauthausen HJ (2006) Evaluation of selected microbial strains and commercial alternative products as seed treatments for the control of *Tilletia tritici*, *Fusarium culmorum*, *Drechslera graminea* and *D. teres*. J Plant Dis Prot 113:150–158

Koch E, Czempinski N, Junge H (2008) Biologische und chemische Charakterisierung der antimikrobiellen Aktivität von *Streptomyces antimycoticus* FZB53. Mitteilungen aus dem Julius Kühn-Institut 417:430–431

Krebs H (2010) Steinbrandbekämpfung im ökologischen Dinkelanbau. Julius-Kühn-Archiv 428:448–449

Kubota M, Hagiwara N, Shirakawa T (2012) Disinfection of seeds of cucurbit crops infested with *Acidovorax citrulli* with dry heat treatment. J Phytopathol 160:364–368

Laterrot H, Pécaut P (1968) Incidence du traitment thermique des semences de tomate sur la transmission du virus de la mosaïque du tabac. Ann Épiphyties 19:159–164

Latin RX, Hopkins DL (1995) Bacterial fruit blotch of watermelon. Plant Dis 79:761–765

Marinelli E, Orzali L, Lotti E, Riccioni L (2012) Activity of some essential oils against pathogenic seed borne fungi on legumes. Asian J Plant Pathol 6:66–74

McGrath MT (2013) Hot water seed treatment protocols. http://vegetablemdonline.ppath.cornell.edu/NewsArticles/HotWaterSeedTreatment.html

Mengulluoglu M, Soylu S (2012) Antibacterial activities of essential oils from medicinal plants against seed-borne bacterial disease agent, *Acidovorax avenae* subsp. *citrulli*. Res Crops 13:641–646

Mohana DC, Prasad P, Vijaykumar V, Raveesha KA (2011) Plant extract effect on seed-borne pathogenic fungi from seeds of paddy grown in Southern India. J Plant Prot Res 51:101–106

Moharam MHA (2010) Biological control of kernel smut of sorghum caused by *Sporisorium sorghi*. PhD thesis, Minia University, Minia, 154 pp

Nagayama K, Watanabe S, Kumakura K, Ichikawa T, Makino T (2007) Development and commercialization of *Trichoderma asperellum* SKT-1 (Ecohope®), a microbial pesticide. J Pestic Sci 32:141–142

Navaratnam SJ, Shuttleworth D, Wallace D (1980) The effect of aerated steam on six seed-borne pathogens. Aust J Exp Agric Anim Hus 20:97–101

Nguefack I, Leth V, Lekagne Dongmo JB, Torp J, Amvam Zollo PH, Nyasse S (2008) Use of three essential oils as seed treatments against seed-borne fungi of rice (Oryza sativa L.). Am-Eurasia J Agric Environ Sci 4:554–560

Nyland G, Goheen AC (1969) Heat therapy of virus diseases of perennial plants. Annu Rev Phytopathol 7:331–354

Pal KK, McSpadden Gardener B (2006) Biological control of plant pathogens. Plant Health Instruct 2:1117–1142. doi:10.1094/PHI-A-2006-1117-02

Pill WG, Collins CM, Goldberger B, Gregory N (2009) Responses of non-primed or primed seeds of 'Marketmore 76' cucumber (*Cucumis sativus* L.) slurry coated with *Trichoderma* species to planting in growth media infested with *Pythium aphanidermatum*. Sci Hortic Amsterdam 121:54–62

Prohens J, Soler S, Nuez F (1999) The effects of thermotherapy and sodium hypochlorite treatments on pepino seed germination, a critical step in breeding programmes. Ann Appl Biol 134:299–305

Qiu J, Westerdahl BB, Giraud D, Anderson CA (1993) Evaluation of hot water treatments for management of *Ditylenchus dipsaci* and fungi in daffodil bulbs. J Nematol 25:686–694

Rane KK, Latin RX (1992) Bacterial fruit blotch of watermelon: association of the pathogen with seed. Plant Dis 76:509–512

Roberti R, Flori P, Pisi A, Brunelli A, Cesari A (2000) Evaluation of biological seed treatment of wheat for the control of seed-borne *Fusarium culmorum*. Z Pflanzenkrankh Pflanzenschutz 107:484–493

Roberts SJ (2009) Evaluation of disinfectants, biological and natural products for control of Brassica black rot (*Xanthomonas campestris* pv. *campestris*). Final report 2008-09. FV 335. HDC, East Malling, UK

Roberts SJ, Amein T, Forsberg G, Kromphardt C, Koch E, Schmitt A, Werner S (2006) Physical and biological seed treatments for control of bacterial diseases of carrots and brassicas caused by *Xanthomonas* spp. In: 11th international conference on plant pathogenic bacteria, Edinburgh, 10–14 July 2006

Roberts S, Brough J, Hunter P (2007) Modelling the spread of *Xanthomonas campestris* pv. *campestris* in module-raised brassica transplants. Plant Pathol 56:391–401

Roberts SJ, Hiltunen LH, Hunter PJ, Brough J (1999) Transmission from seed to seedling and secondary spread of *Xanthomonas campestris* pv. *campestris* in brassica transplants: effects of dose and watering regime. Eur J Plant Pathol 105:879–889

Roberts SJ, Phelps K, Taylor JD, Ridout MS (1993) Design and interpretation of seed health assays. In: Proceedings of the first ISTA plant disease committee symposium on seed health testing, Ottawa, Canada. Agriculture Canada, Ottawa, pp 115–125

Roberts SJ, Ridout MS, Peach L, Brough J (1996) Transmission of pea bacterial blight (*Pseudomonas syringae* pv. *pisi*) from seed to seedling: effects of inoculum dose, inoculation method, temperature and soil moisture. J Appl Bacteriol 81:65–72

Schémaeza B, Somda I, Zida PE, Paco S (2012) Efficacy of plant extracts on *P. sorghina* in seed treatment. World Appl Sci J 20:1549–1553. doi:10.5829/idosi.wasj.2012.20.11.1608

Tegegne G, Pretorius JC (2007) In vitro and in vivo antifungal activity of crude extracts and powdered dry material from Ethiopian wild plants against economically important plant pathogens. Biol Control 52:877–888

Tegegne G, Pretorius JC, Swart WJ (2008) Antifungal properties of *Agapanthus africanus* L. extracts against plant pathogens. Crop Prot 27:1052–1060

Tworkoski T (2002) Herbicide effects of essential oils. Weed Sci 50:425–431

van der Wolf JM, Birnbaum Y, Van Der Zouwen PS, Groot SPC (2008) Disinfection of vegetable seed by treatment with essential oils, organic acids and plant extracts. Seed Sci Technol 36:76–88

Vogelgsang S, Bänziger I, Krebs H, Legro RJ, Sanchez-Sava V, Forrer H-R (2013) Control of *Microdochium majus* in winter wheat with botanicals – from laboratory to the field. Plant Pathol 62:1020–1029

Waldow F, Jahn M (2007) Investigations in the regulation of common bunt (*Tilletia tritici*) of winter wheat with regard to threshold values, cultivar susceptibility and non-chemical protection measures. J Plant Dis Prot 114:269–275

Walker JC (1948) Vegetable seed treatment. Bot Rev 14:588–601

Wang X, Li G, Jiang D, Huang HC (2009) Screening of plant epiphytic yeasts for biocontrol of bacterial fruit blotch (*Acidovorax avenae* subsp. *citrulli*) of hami melon. Biol Control 50:164–171

Williams PH (1967) Occurrence of *Phoma lingam* on cabbage seed from Australia after treatment with hot water. Plant Dis Rep 51:566–569

Wulff EG, Zida E, Torp J, Lund OS (2012) *Yucca schidigera* extract: a potential biofungicide against seed-borne pathogens of sorghum. Plant Pathol 61:331–338

Wunderle J, Leclerque A, Schaffrath U, Slusarenko A, Koch E (2012) Assessment of the loose smut fungi (*Ustilago nuda* and *U. tritici*) in tissues of barley and wheat by fluorescence microscopy and real-time PCR. Eur J Plant Pathol 133:865–875

Yeasmin F, Ashrafuzzamam M, Hossain I (2012) Effects of garlic extract, allamanda leaf extract and Provox-200 on seed borne fungi of rice. The Agriculturists 10:46–50

Yoo M-Y, Cha B, Kim J-C (2013) Recent trends in studies on botanical fungicides in agriculture. Plant Pathol J 29:1–9. doi:10.5423/PPJ.RW.05.2012.0072

Zida EP, Lund OS, Néya JB (2012) Seed treatment with a binary pesticide and aqueous extract of *Eclipta alba* (L.) Hassk. for improving sorghum yield in Burkina Faso. J Trop Agric 50:1–7

Chapter 9
Chemical and Non Chemical Seed Dressing for Leafy Vegetable Crops

M. Lodovica Gullino, Giovanna Gilardi, and Angelo Garibaldi

Abstract Using healthy seeds is a prerequisite in any cropping systems, in order to reduce the further adoption of other disease management strategies in the field during the cultivation. Since seeds are often contaminated, also if at a very low level, by seed-borne pathogens, seed dressing is considered an important method for disease prevention. This is particularly true in the case of seeds of vegetable crops, which very often carry the inoculum of important pathogens. It is well proven with many pathosystems, that a very low percent of infected seeds is able to cause severe losses under greenhouse and field conditions. Seed dressing with chemicals was largely adopted for many decades, because of the availability of effective fungicides, at relatively low cost and the easiness of the treatment. However, recent restrictions in the registration and use of chemicals, stimulated the re-evaluation of old non chemical methods as well as the development of new ones.

This chapter reviews some of the work carried out during the past years in the case of leafy vegetable crops (lettuce, wild and cultivated rocket, lamb's lettuce, chicory, endive, basil, spinach), with chemical and non-chemical measures tested under greenhouse conditions against the most important seed-borne pathogens.

Keywords Seed dressing • Fungicides • Physical methods • Biocontrol agents

1 Introduction

The use of healthy seeds is a prerequisite in any cropping systems, because it permits to strongly reduce the further adoption of other disease management strategies in the field during the cultivation. This is particularly true in the case of seeds of vegetable crops, which very often carry the inoculum of important

M.L. Gullino (✉)
AGROINNOVA, University of Torino, Largo Braccini 2, 10095 Grugliasco, Italy

DISAFA, University of Torino, Largo Braccini 2, 10095 Grugliasco, Italy
e-mail: marialodovica.gullino@unito.it

G. Gilardi • A. Garibaldi
AGROINNOVA, University of Torino, Largo Braccini 2, 10095 Grugliasco, Italy

M.L. Gullino, G. Munkvold (eds.), *Global Perspectives on the Health of Seeds and Plant Propagation Material*, Plant Pathology in the 21st Century 6,
DOI 10.1007/978-94-017-9389-6_9

pathogens. It is well proven with many pathosystems, that a very low percent of infected seeds (lower than 1 %) is able to cause severe losses under greenhouse and field conditions (Gullino et al. 2014).

Since seeds are often contaminated, also if generally at a very low level, by seed-borne pathogens, seed dressing is considered an important method for disease prevention.

After the 1950s, seed dressing has been a very popular strategy for disease control in many crops, including vegetables. The commercialization of very effective fungicides, such as mancozeb and thiram, active against many different pathogens, while being selective for many hosts, made chemical dressing quite widespread. Since such fungicides were able to reduce the presence only of external contaminants, hot water or hot air treatments were necessary to eradicate the presence of pathogens present inside the seeds. The availability, from the late 1960s of systemic fungicides (mainly carboxianilides and benzimidazoles) permitted to overcome the limitations posed by the use of preventative chemicals.

Seed dressing carried out with chemicals remained very popular and largely adopted for many decades, because of the availability of effective fungicides, its relatively low cost and the easiness of the treatment. Indeed, it must be considered that, despite the general trend to reduce the use of chemicals induced by the wide adoption of Integrated Pest Management (IPM) strategies, seed dressing with fungicides has never been considered a high impact practice, since it is carried out in contained environment, using limited amounts of chemicals. Indeed, seed dressing, together with post-harvest treatments, is considered a "minor use" of fungicides (Backhouse 2010).

However, during the past 10 years, the re-registration process imposed by the EU Directive 91/414/EEC, concerning the placing of crop protection products on the market in Europe, the agrochemical portfolio available to European growers has been highly reduced (Leadbeater and Gisi 2010). A number of chemicals, among which many fungicides broadly used for seed dressing, were lost. The new EU regulation for the commercialization of pesticides is further affecting the availability of registered fungicides for minor crops and minor uses (Backhouse 2010). At present, a limited number of chemicals is still available for seed dressing and many of them will probably be lost soon.

In the mean time, the exploitation of organic farming on increasing surfaces, brought to the attention the need to provide farmers with seeds not dressed with chemicals. For quite a few years, the difficulty to find on the market untreated seeds generated problems that were coped with by special temporary permits to use chemically dressed seeds. However, the need of providing healthy seeds, not treated with chemicals, stimulated the re-evaluation of old non chemical methods as well as the development of new ones. Non-chemical seed treatments, carried out by means of physical, micro-organisms and natural products, plant derived products, have been reviewed by Koch and Roberts in this book.

All these reasons stimulate the search for non-chemical methods for seed dressing, while still trying to use in the best way the few remaining fungicides and the European Commission funded two EU projects, under the Vth and VIth framework Programmes, fully devoted to the development of non-chemical seed dressing methods (Nega et al. 2003; Tinivella et al. 2009; Kock et al. 2010; Kock and Roberts 2014).

This chapter will review some of the work carried out during the past years in the case of leafy vegetable crops (lettuce, wild and cultivated rocket, lamb's lettuce, chicory, endive, basil, spinach), with chemical and non-chemical measures tested under greenhouse conditions against the most important seed-borne pathogens.

2 Soilborne Pathogens

Vascular wilts, caused by *Fusarium oxysporum* and *Verticillium dahliae* are important on most leafy vegetables, causing severe losses. Fusarium wilts on salad crops, basil and spinach have been reviewed respectively by Matheron and Gullino (2012), Gullino et al. (2012) and by Correll et al. (1994). Since the different *formae speciales* of *Fusarium oxysporum* and *V. dahliae* causing wilts on the different crops are often seed-transmitted (Gullino et al. 2014), seed dressing is considered one of the most important preventative treatment.

In the case of lettuce, among chemicals, prochloraz, mancozeb, carbendazim resulted very effective, providing, in the presence of a high level of seed infection in the untreated control, consistent results against *Fusarium oxysporum* f. sp. *lactucae*, in the different trials carried out (Tables 9.1, 9.2, and 9.3). Thiram and acibenzolar-*S*-methyl were less effective, but still provided satisfactory disease control (Tables 9.2 and 9.3) (Gilardi et al. 2005; Lopez-Reyes et al. 2014a, b).

Among registered biocontrol agents, *Bacillus subtilis* QST 713 gave interesting and consistent disease reduction (Tables 9.1 and 9.2), while *B. subtilis* FZB24 and MB1 600 were much less effective (Table 9.1). A very low efficacy was shown by *Streptomyces griseoviridis* K61 (Table 9.1). Very interesting results, with high disease reduction were provided by the antagonistic strains of *Fusarium oxysporum* MSA35 and 251/2 (Tables 9.1 and 9.2) and by several isolates of *Pseudomonas* spp. (Table 9.2). Interesting results were also obtained with the use of several mixtures of microorganisms (*Trichoderma harzianum* ICC 012 + *Trichoderma viride* ICC080; *Glomus* spp. + *Bacillus megaterium* + *Trichoderma* 1010 (Table 9.2) (Gilardi et al. 2005; Lopez-Reyes et al. 2014a).

Among natural products, thyme and savory essential oils, applied at 1 % as spray, were very effective, providing a disease reduction higher than 75 % (Table 9.3) (Lopez-Reyes et al. 2014b).

Also hot water treatments (50 °C for 10 min) provided interesting results. The level of control of such treatment is comparable with that of the best chemicals and is not improved when combined with fungicides or biocontrol agents (Fig. 9.1) (Lopez-Reyes et al., personal communication).

On basil, seed dressing with prochloraz, thiram, savory and thyme essential oil provided statistically similar results, determining a 60–70 % reduction of Fusarium

Table 9.1 Efficacy of microorganisms and different fungicides for dressing seeds of lettuce against *Fusarium oxysporum* f. sp. *lactucae*. Data are average of six trials expressed as efficacy compared to the inoculated and non-treated control 23–37 days after sowing (From Gilardi et al. 2005)

Treatments	Dosage a. i./kg of seed	Seed application	% of disease reduction	
Inoculated and non-treated control	–	–	0.0 (31.4 %)[a]	f
Bacillus subtilis QST 713	0.1 g	Spray	42.7	b-f
Bacillus subtilis FZB24	0.1 g	Spray	24.7	ef
Bacillus subtilis MB1 600	0.1 g	Spray	33.4	d-f
Streptomyces griseoviridis K61	0.02 g	Spray	28.8	ef
Pseudomonas chlororaphis MA342	0.3 ml	Spray	26.9	ef
Fusarium oxysporum 251/2	1×10^7 CFU[b]	Spray	47.8	a-e
Fusarium oxysporum 251/2	1×10^7 CFU	Mixing of dry powder	38.7	c-f
Fusarium oxysporum MSA 35	1×10^7 CFU	Spray	59.9	a-e
Fusarium oxysporum MSA 35	1×10^7 CFU	Mixing of dry powder	76.8	a-d
Mancozeb	4.8 g	Spray	84.3	a-c
Carbendazim	1.3 g	Spray	91.2	a
Prochloraz	0.9 g	Spray	86.7	ab
Thiram	1.5 g	Spray	44.1	b-f

Seed sample was naturally contaminated by *Fusarium oxysporum* f. sp. *lactucae* at 0.9 %
The mean values of the same column followed by the same letter do not differ significantly according to Tukey test ($p = 0.05$)
[a]% of contaminated seeds in the inoculated non-treated control
[b]*CFU* colony forming units

wilt, incited by *F. oxysporum* f. sp. *basilici* (Table 9.4) (Lopez-Reyes et al. 2014b). The same level of efficacy was previously consistently shown by the antagonistic strain of *F. oxysporum* 251/2 (Table 9.5) (Garibaldi et al. 1997). Hot air treatment (65 °C for 10 min) provided a partial control against this pathogen; the level of efficacy provided by this physical treatment improved when applied in combination with savory essential oil (Fig. 9.2) (Lopez-Reyes et al., personal communication).

Hot water treatments and chlorine were tested for eradication of *Verticillium dahliae* as well as *Cladosporium variabile* and *Stemphylium botryosum*, three seed-borne pathogens of spinach. Seed treatments with chlorine, at 1.2 %, for 10–40 min or hot water (at 40–60 °C) for 10–40 min, showed different degrees of efficacy against this three pathogens of spinach (Fig. 9.3) (Du Toit et al. 2005). Chlorine

Table 9.2 Efficacy of different products applied as powder and spray seed treatments of artificially inoculated lettuce seeds with *Fusarium oxysporum* f. sp. *lactucae*, race 1. Data are average of four trials expressed as efficacy compared to the inoculated and non-treated control 40 days after sowing (From Lopez-Reyes et al. 2014a)

Treatments	Dosage a. i./kg of seeds	Seed application	% of disease reduction	
Inoculated and non-treated control	–	–	0.0 (27.7)[a]	c
Acibenzolar-*S*-methyl	0.1 g	Spray	59.2	ab
Prochloraz	1.0 g	Spray	91.6	a
Thiram	9.8 g	Spray	71.8	ab
Bacillus subtilis – QST 713	10.0 g	Spray	55.9	ab
Bacillus subtilis BA41; *Streptomyces* sp. SB15; *Trichoderma harzianum* TH02; *Pseudomonas proradix10*; *Glomus caledonium* GM24; *Glomus coronatum* GU53; *Gladius intraradices* GB67 *Trichoderma* spp.	2.0 g	Mixing of dry powder	59.2	ab
Streptomices griseoviridis K61	8.0 g	Spray	35.3	b
Streptomyces spp. SB14; *Glomus coronatum* GO01; *Glomuscoronatum* GU53; *Glomus caledonium* GM24; *Bacillus subtilis* SR63; *Pseudomonas* spp. PM46; *Ulocladium* spp. UO18	2.0 g	Spray	40.7	b
Trichoderma harzianum ICC 012 + *Trichoderma viride* ICC 080	2.0 g	Spray	66.1	ab
Glomus spp. 5 % + *Bacillus megaterium* 10^4 CFUg^{-1} + *Trichoderma* 10^{10} CFUg-^{1}	2.0 g	Spray	59.2	ab
Pseudomonas sp. FC6B (EU836173)	1×10^7 CFU [b]	Spray	57.8	ab
Pseudomonas putida FC7B (EU836174)	1×10^7 CFU	Spray	69.7	ab
Pseudomonas sp. FC8B (EU836171)	1×10^7 CFU	Spray	66.1	ab
Pseudomonas sp. FC9B (EU836172)	1×10^7 CFU	Spray	76.1	ab
Pseudomonas sp. FC24B (EU836173)	1×10^7 CFU	Spray	53.8	ab
Fusarium oxysporum 251/2	1×10^7 CFU	Mixing of talc formulation	61.7	ab
Fusarium oxysporum MSA35	1×10^7 CFU	Mixing of talc formulation	52.3	ab

Seed sample was naturally contaminated by *Fusarium oxysporum* f. sp. *lactucae* at 1.1 %
The mean values of the same column followed by the same letter do not differ significantly according to Duncan test ($p = 0.05$)
[a]% of contaminated seeds in the Inoculated non-treated control
[b]CFU colony forming units

treatment, at 1.2 %, for 10–40 min was very effective against *C. variabile* and *V. dahliae*, while at least 60 min of treatment were needed against *S. botryosum*. The treatment of spinach seeds did not have a significant effect on seed quality, even after 60 min in 1.2 % NaOCl (Du Toit et al. 2005). Hot water treatment of

Table 9.3 Efficacy of different products applied as powder and spray seed treatments of artificially inoculated lettuce seeds with *Fusarium oxysporum* f. sp. *lactucae*. Data are average of four trials expressed as efficacy compared to the inoculated and non-treated control 40 days after sowing (From Lopez-Reyes et al. 2014b)

Treatments	Dosages a. i./kg of seeds	Seed application	% of disease reduction	
Inoculated and non-treated control	–	–	0.0 (24.0)[a]	c
Acibenzolar-*S*-methyl	0.1 g	Spray	57.9	ab
Prochloraz	1 g	Spray	90.0	a
Thiram	9.8 g	Spray	70.9	ab
Satureja montana (Savory essential oil)	0.1 %	Spray	49.8	a-c
Satureja montana (Savory essential oil)	1 %	Spray	78.3	ab
Thymus vulgaris (Thyme essential oil)	0.1 %	Spray	34.1	bc
Thymus vulgaris (Thyme essential oil)	1 %	Spray	77.0	ab

Seed sample was naturally contaminated by *Fusarium oxysporum* f. sp. *lactucae* at 1.1 %
The mean values of the same column followed by the same letter do not differ significantly according to Duncan test ($p = 0.05$)
[a]% of contaminated seeds in the inoculated non-treated control

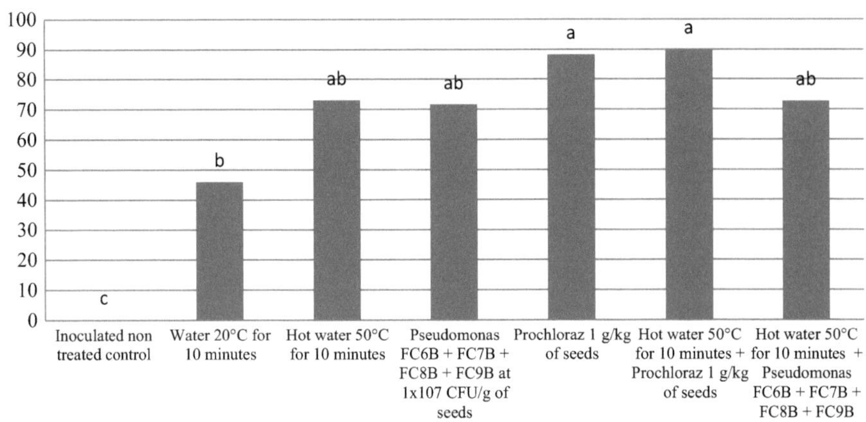

Fig. 9.1 Efficacy of physical, chemical, biological seed treatments used alone and combined against artificially inoculated seeds with *Fusarium oxysporum* f. sp. *lactucae*. Data are average of three trials expressed as efficacy compared to the inoculated and non-treated control 50 days after sowing. Seed sample was naturally contaminated by *Fusarium oxysporum* f. sp. *lactucae* at 11.6 %. The mean values followed by the same letter do not differ significantly according to Duncan test ($p = 0.05$) (From Lopez-Reyes et al. 2014a, b)

spinach seeds provided less interesting results against *V. dahliae*, since the temperature required for its eradication (≥50 °C) negatively affect seed germination (Table 9.6) (Du Toit et al. 2005). Recent researches carried out by Maruthachalam et al. (2013) were aimed to better understand the localization of the pathogen in spinach seeds, showing that the maximum concentration of *V. dahliae* was in the

Table 9.4 Efficacy of different products applied as powder and spray seed treatments of artificially inoculated lettuce seeds with *Fusarium oxysporum* f. sp. *basilici*. Data are average of four trials expressed as efficacy compared to the inoculated and non-treated control 40 days after sowing (From Lopez-Reyes et al. 2014b)

Treatments	Dosages a. i./kg of seeds	Seed application	% of disease reduction	
Inoculated and non-treated control	–	–	0.0 (27.2)[a]	b
Prochloraz	1 g	Dry powder	59.6	a
Thiram	9.8 g	Dry powder	70.0	a
Satureja montana (Savory essential oil)	10 %	Fumigation	66.1	a
Thymus vulgaris (Thyme essential oil)	10 %	Fumigation	65.1	a

Seed sample was naturally contaminated by *Fusarium oxysporum* f. sp. *basilici* at 8.7 %
The mean values of the same column followed by the same letter do not differ significantly according to Duncan test ($p = 0.05$)
[a]% of contaminated seeds in the inoculated non-treated control

Table 9.5 Effect of seed treatment integrated with soil drench treatment with ipovirulent *Fusarium oxysporum* on Fusarium wilt of basil (cv. Genovese gigante). Data are expressed as disease reduction compared to the non-treated control (From Garibaldi et al. 1997)

			% of disease reduction of		
Treatments	Dosage a. i	Application	Trial 1	Trial 2	Trial 3
Non-treated control	–	–	0.0 (25 %)[a]	0.0 (46.2 %)	0.0 (13.8 %)
Fusarium oxysporum 251/2	3×10^7 CFU[b] per g of seed + 10^5/ml of soil	Spray seed dressing + soil drench	87.2	69.5	71.0

Soil was artificially inoculated with 1×10^3 CFU/ml of a talc formulation of *Fusarium oxysporum* f. sp. *basilici*
[a]% of contaminated seeds in the non-treated control
[b]CFU colony forming units

seed coat. This findings will help in the development of more effective seed treatments.

Among the tested microorganisms, *Streptomyces griseoviridis* reduced the contamination of spinach seeds from *S. botryosum*, *Verticillium* and *Alternaria* spp. respectively by 88 %, 74 % and 84 % respectively, while it was not effective against *Fusarium* spp.. *Bacillus pumilis* showed some efficacy against *Fusarium* spp.. Thiabendazole was highly effective against *Verticillium* spp. and *Fusarium* spp. (Table 9.7) (Cummings et al. 2009).

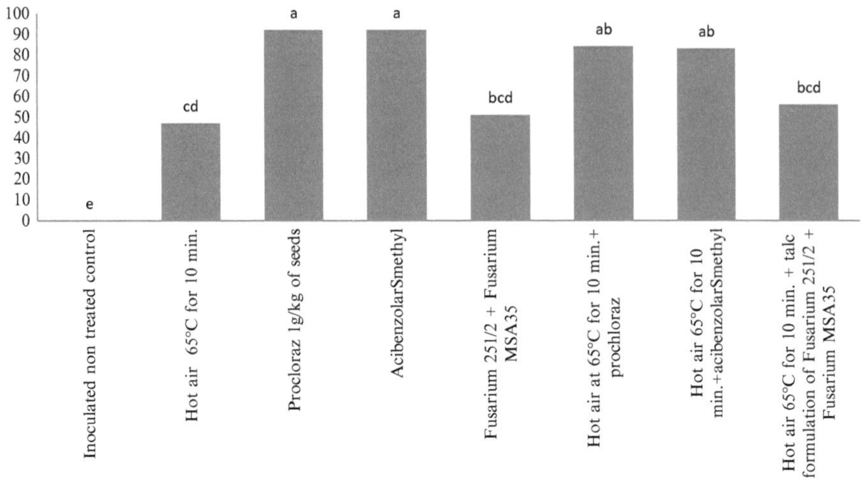

Fig. 9.2 Efficacy of physical, chemical, biological seed treatments used alone and combined against artificially inoculated seeds with *Fusarium oxysporum* f. sp. *basilici*. Data are average of three trials expressed as efficacy compared to the inoculated and non-treated control 40 days after sowing. Seed sample was naturally contaminated by *Fusarium oxysporum* f. sp. *basilici* at 48.4 % The mean values followed by the same letter do not differ significantly according to Duncan test ($p = 0.05$) (From Lopez-Reyes et al., personal communication)

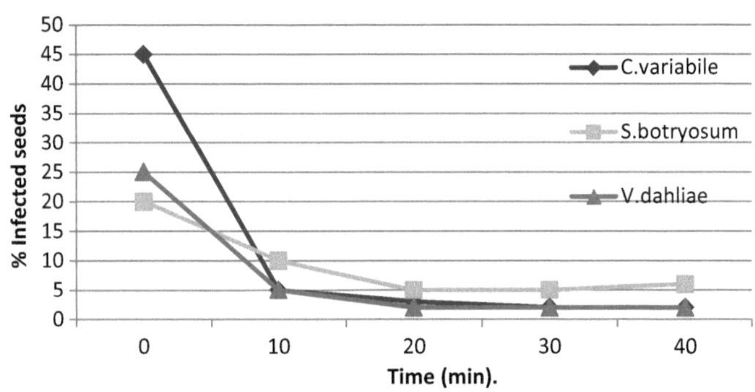

Fig. 9.3 Effect of sodium hypochlorite at 1.2 % for 10, 20, 30 and 40 min against *Cladosporium variable*, *Stemphylium botryosum* and *Verticillium dahliae* on seeds of spinach. Data expressed as efficacy compared to the inoculated and non-treated (From du Toit et al. 2005)

3 Foliar Pathogens

Since several foliar pathogens are seed-borne, they can be properly managed by seed treatment.

The good results shown by the treatment of spinach seeds with 1.2 % NaOCl against *C. variabile* and, with at least 60 min of treatment against *S. botryosum* have

Table 9.6 Hot water treatment recommended for the eradication of seed-borne pathogens of spinach (From du Toit et al. 2005)

Pathogen	Temperature and duration of application[a]
Cladosporium variabile	40 °C for 10 min
Verticillium dahliae	45 °C for 20 min
Stemphylium botryosum	55 °C for 20 min
Cladosporium variabile, *Stemphylium botryosum*, *Verticillium dahliae*	50 °C for 20 min

[a]Strong reduction of seed germination with the treatments: 50 °C ≥ 30 min; 55 °C and 60 °C ≥10 min

Table 9.7 Effectiveness of different seed treatments with micro-organisms and fungicide against several pathogens of spinach (From Cummings et al. 2009)

Trattamento	Dosage a.i./kg of seeds	*Stemphylium botryosum*	*Verticillium* spp.	*Alternaria* spp.	*Fusarium* spp.
Non-treated control	–	0.0 (4.0)[a]	0.0 (49.7)	0.0 (13.3)	0.0 (2.2)
Bacillus subtilis	0.0043 g	12.5	2.5	23.0	23.0
Streptomyces griseoviridis	0.25 g	88.0	74.2	84.9	0.0
Bacillus subtilis	0.0043 g	50.0	9.1	45.7	0.0
Bacillus pumilis	0.0002 g	25.0	0.0	11.3	50.0
Trichoderma harzianum T22	0.029 g	56.0	17.0	22.0	0.0
Thiabendazole	0.517 g	0.0	98.0	13.0	100.0

[a]% of contaminated seeds in the non-treated control

already been mentioned (Du Toit et al. 2005). Hot water treatments of 40 °C for 10 min, were effective against *Cladosporium variabile* on spinach, while higher temperatures, resulting phytotoxic were needed against *Stemphylium botryosum* (Table 9.6) (Cummings et al. 2009).

Among the different microorganisms tested, *Bacillus subtilis* was effective against *S. botryosum* and *Alternaria* spp., while the fungicide tested, thiabendazole, was not effective (Table 9.7) (Cummings et al. 2009).

In the case of *Phoma valerianellae*, a pathogen causing a leaf spot of lamb's lettuce which is transmitted through infected seeds, also in the presence of a very high level of seed infection, treatments with aerated steam (2 min) or hot water (50 °C for 30 min or 53 °C for 10 min) were very effective in eradicating the pathogen. Satisfactory results were also provided by treatments with electrons and thyme oil, while *Streptomyces griseoviridis* K61 was not effective (Table 9.8) (Schmitt et al. 2009).

Table 9.8 Efficacy of physical, chemical and biological treatments of seeds of corn salad naturally infested with *Phoma valerianellae* (From Schmitt et al. 2009)

		% of disease reduction on	
Treatments	Dosage a. i./kg of seeds	Sample 1	Sample 2
Non-treated control	–	0 (65)[a]	0 (14)
Aerated steam (2 min)	–	92	71
Aerated steam (5 min)	–	85	35
Hot water (50 °C, 30 min)	–	92	57
Hot water (53 °C, 10 min)	–	87	78
Electrons (110 KV/12 kGy)	–	46	70
Electrons (110 KV/24 kGy)	–	53	71
Thyme oil (0.1 %)	–	41	71
Streptomyces griseoviridis k61	5 g	15	35
Thiram	0.67 g	15	71

[a]% of contaminated seeds in the non-treated control

Conclusions

Several research groups did concentrate their research during the past years on the exploitation of non-chemical control methods for seed dressing, under the pressure determined by the loss of several effective chemicals as well as by the strong request coming from organic farmers, needing healthy seeds, not treated with chemicals.

The different researches carried out show that many options are available, which provide interesting results. Some of them, such as hot water or aerated steam treatments offer an efficacy comparable to that provide by fungicides, while biocontrol agents are often less effective and, even more important, not always offer consistent results in different trials.

The use of essential oils needs to be further exploited. Actually, with some pathogens able to infect the seeds internally, there is a need of products able to penetrate the seeds. Hot water and some essential oils seem to have the capability of acting not only externally.

Much more research is needed in order to optimize the use of non-chemical methods and different approaches must be developed with different pathosystems. Moreover, the adoption of non-chemical measures requires very well trained technicians able to provide adequate assistance to growers, often reluctant to adopt techniques appearing not easy to apply.

However, the present regulatory situation and future losses of chemicals at present available, coupled with an increasingly importance of seed-borne pathogens will force growers to rely more and more of alternative measures.

All the researches carried out by different groups show that the results achieved can be quite interesting and that alternative methods, if well applied, can be very competitive with old chemicals.

Acknowledgements Part of the research leading to some of the results reported has received funding from the European Union's Seventh Framework Programme (FP7/2007–2013) under grant agreement n° 311875 "TESTA – Seed health: development of seed treatment methods, evidence for seed transmission and assessment of seed health".

References

Backhouse GE (2010) Regulatory aspects in chemical control of fungal diseases: impact on efficient plant production. In: Gisi U, Chet I, Gullino ML (eds) Recent developments in management of plant diseases. Springer, Dordrecht, pp 47–55

Correll JC, Morelock TE, Black MC, Koike ST, Brandenberger LP, Dainello FJ (1994) Economically important diseases of spinach. Plant Dis 78:653–660

Cummings JA, Miles CA, du Toit LJ (2009) Greenhouse evaluation of seed and drench treatments for organic management of soilborne pathogens of spinach. Plant Dis 93:1281–1292

Du Toit LJ, Derie ML, Hernandez-Perez P (2005) Verticillium wilt in spinach seed production. Plant Dis 89:4–11

Garibaldi A, Gullino ML, Minuto G (1997) Diseases of basil and their management. Plant Dis 81:124–132

Gilardi G, Tinivella F, Gullino ML, Garibaldi A (2005) Seed dressing to control *Fusarium oxysporum* f. sp. *lactucae*. J Plant Dis Prot 112:240–246

Gullino ML, Gilardi G, Garibaldi A (2014) Seed-borne fungal pathogens of leafy vegetable crops. In: Gullino ML, Munkvold G (eds) Global perspectives on the health of seeds and plant propagation material. Springer, Dordrecht, pp 47–56

Gullino ML, Katan J, Garibaldi A (2012) Fusarium wilt of sweet basil. In: Gullino ML, Katan J, Garibaldi A (eds) Fusarium wilts of greenhouse crops. APS Press, St Paul, pp 185–190

Kock E, Roberts SE (2014) Non-chemical seed treatment in the control of seed-borne pathogens. In: Gullino ML, Munkvold G (eds) Global perspectives on the health of seeds and plant propagation material. Springer, Dordrecht, pp 105–123

Kock E, Schmitt A, Spephan D, Kromphardt C, Jahn M, Krauthausen HJ, Forsberg G, Werner S, Amein T, Wright SAI, Tinivella F, Gullino ML, Roberts SJ, van der Wolf J, Groot SPC (2010) Evaluation of non-chemical seed treatment methods for the control of *Alternaria dauci* and *A. radicina* on carrot seeds. Eur J Plant Pathol 127:99–112

Leadbeater A, Gisi U (2010) The challenges of chemical control of plant diseases. In: Gisi U, Chet I, Gullino ML (eds) Recent developments in management of plant diseases. Springer, Dordrecht, pp 3–17

Lopez-Reyes JG, Gilardi G, Garibaldi A, Gullino ML (2014a) Efficacy of bacterial and fungal biocontrol agents as seed treatments against *Fusarium oxysporum* f. sp. *lactucae*. J Phytopathol (accepted)

Lopez-Reyes JG, Gilardi G, Garibaldi A, Gullino ML (2014b) Efficacy of essential oils as seed treatments against *Fusarium oxysporum* f. sp. *basilici* on basil and *Fusarium oxysporum* f. sp. *lactucae* on lettuce. J Essent Oil Res (Submitted)

Maruthachalam K, Klosterman SJ, Ancheta A, Mou BQ, Subbarao KV (2013) Colonization of spinach by *Verticillium dahliae* and effects of pathogen localization on the efficacy of seed treatments. Phytopathology 103:268–280

Matheron M, Gullino ML (2012) Fusarium wilts of lettuce and other salad crops. In: Gullino ML, Katan J, Garibaldi A (eds) Fusarium wilts of greenhouse crops. APS Press, St Paul, pp 175–183

Nega E, Ulrich R, Werner S, Jahn M (2003) Hot water treatment of vegetable seeds. An alternative seed treatment method to control seed-borne pathogens in organic farming. J Plant Dis Prot 110:220–234

Schmitt A, Koch E, Stephan D, Kromphardt C, Jahn M, Krauthausen HJ, Forsberg G, Werner S, Amein T, Wright SAI, Tinivella F, van derWolf J, Groot SPC (2009) Evaluation of non-chemical seed treatment methods for the control of *Phoma valerianellae* on lamb's lettuce seeds. J Plant Dis Prot 116:200–207

Tinivella F, Hirata LM, Celan MA, Wright SAI, Amein T, Schmitt A, Koch E, van Der JM, Groot SPC, Stephan D, Garibaldi A, Gullino ML (2009) Control of seed-borne pathogens on legumes by microbial and other alternative seed treatments. Eur J Plant Pathol 123:139–151

MIX
Papier aus verantwortungsvollen Quellen
Paper from responsible sources
FSC® C105338

If you have any concerns about our products,
you can contact us on
ProductSafety@springernature.com

In case Publisher is established outside the EU,
the EU authorized representative is:
Springer Nature Customer Service Center GmbH
Europaplatz 3, 69115 Heidelberg, Germany

Printed by Libri Plureos GmbH
in Hamburg, Germany